INSECTS OF NORTH AMERICA

FALCONGUIDES®

INSECTS OF NORTH AMERICA

A FIELD GUIDE TO OVER 300 INSECTS

DAVID M. PHILLIPS PhD

FALCON GUIDES

GUILFORD, CONNECTICUT

FALCONGUIDES®

An imprint of The Rowman & Littlefield Publishing Group, Inc.
4501 Forbes Blvd., Ste. 200
Lanham, MD 20706
www.rowman.com
Falcon and FalconGuides are registered trademarks and Make Adventure Your Story is a trademark of The Rowman & Littlefield Publishing Group, Inc.

Distributed by NATIONAL BOOK NETWORK

British Library Cataloguing in Publication Information available

Library of Congress Cataloging-in-Publication Data
Names: Phillips, David M., author.
Title: Insects of North America : a field guide / David M. Phillips, PhD.
Description: Guilford, Connecticut : FalconGuides, [2019] | Includes index. |
Identifiers: LCCN 2018053816 (print) | LCCN 2018055852 (ebook) | ISBN 9781493039241 (e-book) | ISBN 9781493039234 (paperback) | ISBN
Subjects: LCSH: Insects—North America—Identification.
Classification: LCC QL473 (ebook) | LCC QL473 .P45 2019 (print) | DDC 595.7097—dc23
LC record available at https://lccn.loc.gov/2018053816

The paper used in this publication meets the minimum requirements of American National Standard for Information Sciences—Permanence of Paper for Printed Library Materials, ANSI/NISO Z39.48-1992.

Printed in the United States of America

CONTENTS

Acknowledgments ... vii
Introduction ... viii
Photographing Insects ... xvi
Nomenclature ... xviii

1 Springtails ... 1
2 Silverfish ... 3
3 Earwigs ... 5
4 Termites ... 7
5 Lice ... 10
6 Fleas ... 13
7 Barklice and Booklice ... 15
8 Psyllids ... 17
9 Thrips ... 19
10 Scorpionflies ... 21
11 Dragonflies ... 23
12 Damselflies ... 28
13 Mayflies ... 33
14 Caddisflies ... 37
15 Cockroaches ... 40
16 Mantises ... 44
17 Stick Insects ... 47
18 Grasshoppers ... 49
19 Crickets ... 58
20 Hemiptera ... 63
21 Beetles ... 100
22 Neuroptera ... 151
23 Butterflies ... 160
24 Moths ... 175
25 Flies ... 213

26 Wasps 251
27 Bees 275
28 Ants 285

Glossary 289
Index 293
About the Author 299

ACKNOWLEDGMENTS

My wife, Robin Maguire, encouraged me to write this book. She also made my awkward English readable. She corrected the grammar and organized the figures. I could not have written the book without her help and support.

INTRODUCTION

There are about 90,000 species of insects in North America, but don't worry, I am not going to describe them all. Instead, I have selected what I feel are some of the most common and interesting North American insects. Some are particularly interesting because they are pests, prey on pests, or transmit diseases and others because of their unusual lifestyles or how they interact with other animals or plants. In the discussions of each insect, I concentrate on the intriguing aspects of their life rather than the details of their anatomy. Unless you are an expert with lots of training and experience and have access to a microscope and the use of a key to the particular group of insects, you will most likely identify insects by comparing them to a photograph. Therefore, I included at least one photograph of each insect. As much as feasible, I have avoided technical terms, although some terms are needed in order to discuss insects and their lives.

Insects can be found almost everywhere in North America. Some make their home in fields, others in forests. Many species live in leaf litter or under the ground. There are aquatic insects and those that live in deserts. The good news is that in order to observe or photograph many kinds of these little animals, it is not necessary to crisscross North America because many insect species can be found throughout much of the continent. However, the Rocky Mountains and deserts are formidable barriers for many insects. In fact, many kinds of insects are primarily, or only, found east of the Rockies. Insects are the only invertebrates that can fly. This has allowed them to escape predators and find food and new territories.

Insect Development

Insects belong to a group of animals called arthropods, which includes lobsters, crabs, spiders, and a few other kinds of animals. Arthropods are characterized by an external skeleton, or exoskeleton, and jointed appendages.

Most insects develop in one of two ways. Primitive insects are those that are most similar to ancestral insects. The young of the primitive insects like grasshoppers, cockroaches, and bedbugs look like the adults, although their proportions are generally a little different. They are wingless and do not have mature sexual organs. Their development is referred to as simple or incomplete morphogenesis. The young are called nymphs, unless they are aquatic, and then they are usually referred to as naiads.

On the other hand, the young of the more modern (derivative) insects like flies, bees, and butterflies look nothing like they will when they become adults. These young are usually called larvae, although larvae of some insects are referred to as grubs, caterpillars, or maggots. When larvae are old enough, they enter a dormant pupal stage where they undergo radical transformation as they mature into adults. Their development is referred to as complete metamorphosis.

Typically, insect nymphs and larvae grow rapidly. Some insects that undergo complete metamorphosis grow at an amazing rate. There are, for example, caterpillars that increase in size 1,000 times in a week and flies that can complete their entire life cycle from microscopic egg to adult in little more than a week. There are exceptions: Some beetle larvae can live inside of trees for years, and some periodical cicadas take 17 years to mature. Because insects are cold-blooded, nymphs and larvae grow faster when they are warm.

Surviving the Winter

Although cold winters take a toll on many North American animals, including insects, most insects that live in temperate climates can cope with temperatures well below 32 degrees Fahrenheit at some stage of their life. Each species of insect overwinters at just one stage. Many insect species overwinter as eggs, but some species spend the winter as nymphs, larvae, pupae, or adults. It is usually only this stage that can survive the winter. Although the stage that overwinters can survive a typical winter or even an exceptionally cold winter, many insects die when a sudden cold wave occurs in the spring after they have overwintered and progressed to the next stage of their development.

Regardless of the stage of development, because they are "cold-blooded," the body temperature of an insect is usually the same as the air, plant, or earth around it. A few kinds of insects are able to fly long distances to warmer climates in the winter, and some species can dig below the frost line, but most insects cannot escape winter.

In the fall some insects crawl below the ground or leaf litter or lay their eggs there. Others may overwinter inside of plants or trees. Although the ground, leaf litter, and trees provide some protection from the cold, when the temperature falls significantly below 32 degrees Fahrenheit, the temperature in plants, trees, and leaf litter also falls below freezing. Of course, trees do not freeze to death, nor do insects. When spring comes, leaves and plants emerge, and insects come forth from everywhere because animals and plants that live in temperate climates have ways to cope with freezing temperatures.

Some kinds of insects can survive even though their tissues freeze. However, most insects that live in temperate climates survive the winter because their tissues do not form ice crystals when the temperature falls below freezing. The blood (hemolymph) and cells of insects contain a small molecule called glycerol as well as certain proteins, sugars, and salts that act as antifreeze to prevent ice formation. In fact, some kinds of North American insects are active even when their temperature dips considerably below freezing.

The Exoskeleton

The bodies of adult insects have three parts: the head, thorax, and abdomen. The head contains the antennae, mouth, and eyes. Wings and legs are situated in the thorax, and the abdomen contains the reproductive system and most of the digestive and respiratory systems.

The exoskeleton of the head is generally very hard. On the other hand, the exoskeleton of the abdomen is softer because it needs to enlarge to accommodate food and, in the case of females, grow eggs. In a similar way, the exoskeleton of the different segments of an insect's legs is usually very hard, whereas the exoskeleton of the leg joints is soft, flexible, and contains an elastic protein called resilin, which allows the joints to stretch and compress as the insect walks. Many of the parts of the exoskeletons of insects and those of other arthropods are tube-shaped because a tube is the strongest way to construct a support with a given amount of material, which is why the frames of bicycles and racing cars are constructed of tubes.

The exoskeleton protects the underlying tissues, and muscles are attached to it. It is composed of proteins, especially proteins called sclerotin and polysaccharides called chitins. Polysaccharides are molecules composed of many sugars linked together. The exoskeleton is a light and exceptionally strong substance for its weight. It can be either exceptionally hard or soft and flexible, depending on the different types of chitins and proteins that it is composed of and how they are linked together.

One can think of the exoskeleton as somewhat analogous to a suit of armor. In insects, the individual pieces of the exoskeleton are called sclerites, and just as the plates in a suit of armor have names, the sclerites have names. The names of some of these sclerites are particularly useful to know when identifying insects. One of the most important sclerites in this regard is the one that is on top of the thorax, which is called the pronotum. Another important sclerite is the first segment of insect legs, the femur. Although some entomologists need to be familiar with the names of almost all the sclerites as well as the wing veins, it is beyond the scope of this book.

Because it is composed of nonliving material, the exoskeleton does not grow. In order to enlarge, insects make a new exoskeleton under the old one. They then take in air to enlarge and shed the old exoskeleton. At first, the new exoskeleton is very soft, but it quickly hardens and darkens by a chemical reaction similar to the tanning of leather.

Replacing the old exoskeleton with the new, larger one is called molting. Insects generally undergo several molts before their final adult molt. Three to five molts are typical of most species; however, some kinds of insects undergo a dozen or more molts. The nymph or larva that emerges from an egg is referred to as a first instar nymph or larva. After it molts again, it becomes a second instar nymph or larva, and so on. When they reach adulthood, insects cease to molt and thus do not increase in size.

Although adult insects cannot increase in size, due to genetic variation and differences in the nutrition when they are young, some adults of a species may be larger than others. Females are larger than males in many kinds of insects. This is because the larger females need to accommodate eggs.

Retaining Water

The exoskeleton of insects is unique among arthropods because it is coated with wax that renders it totally waterproof and prevents water from evaporating from an insect's body. The exoskeletons of other land arthropods like spiders and centipedes usually lack wax coatings. Even insects that are the size of a grain of sand are covered with a minuscule wax coat. Although the wax that coats the insect exoskeleton may be only one molecule thick, it acts as a very effective water barrier.

Small land animals have a much more difficult problem with conserving water than large animals because their surface area is relatively large compared to their size. Consider a small insect flying on a hot, dry, sunny day. A drop of water the size of the insect would evaporate in a short time under these conditions, but the little insect flies on. Although other small land animals face this problem, none can avoid dehydration as well as insects. Thus, little animals like worms and slugs must live in humid places, or they will dry up and die. Even land animals that are related to insects like spiders and scorpions cannot retain water as effectively as insects, which is one reason why most of them are nocturnal. The ability to avoid drying out has enabled some insects to fly on sunny days and thrive in the open fields and even the driest deserts.

Because vertebrates, especially mammals and birds, have high metabolisms, their respiratory system provides for the exchange of a large amount of oxygen and carbon dioxide. However, most mammals and birds lose a lot of water when they breathe because the air that is exhaled absorbs water from the moist surfaces of their lungs. Insects have a vastly different respiratory system. In some of the very tiniest insects, diffusion of oxygen and carbon dioxide through their exoskeleton is sufficient. However, all but the very smallest insects have a respiratory system.

Vertebrates transport oxygen in their blood, but the blood (hemolymph) of insects is not involved in respiration. Instead of lungs, insects have openings in their thorax and abdomen that lead to a branching network of air-filled, exoskeleton-lined tubes called trachea that end in minute, usually fluid-filled tubes called tracheoles that transport oxygen to, and carbon dioxide from, their tissues. Insects that need a lot of exchange of oxygen and carbon dioxide have enlarged areas in their trachea called air sacs, which expand and contract to increase the flow of air in their trachea. Valves, called spiracles, at the openings of the trachea control the size of the opening. When an insect is inactive, spiracles close to prevent evaporation. In small insects, diffusion is sufficient to transport enough air into the trachea, but many larger insects breathe by expanding and contracting their abdomen.

Most vertebrates lose a considerable amount of water in their urine and feces. The insect digestive and urinary systems are designed to conserve water. In insects, the portions of their digestive tract nearest the mouth, called the foregut, and nearest the anus, termed the hindgut, are lined with exoskeleton that has the same wax coating as the insect's outer exoskeleton. Digestion takes place exclusively in the middle, or midgut, of the digestive tract, which is devoid of exoskeleton. Malpighian tubules, the equivalent of vertebrate kidneys, empty into the midgut. Insects lose relatively little water from their feces because of glands in their midgut or hindgut or their Malpighian tubules absorb the water.

In addition to conserving water, insects and other animals obtain "metabolic water" by metabolizing fats and carbohydrates. This enables some insects to survive for long periods without drinking. In fact, there are even species that can live exclusively on metabolic water.

Size Matters

We tend to be most familiar with the relatively large insects, because they are the easiest to see, but most insects are very small. Some are only the size of a grain of rice, while others are too small to see with the naked eye. However, even the relatively large insects are small animals. Twenty-thousand houseflies weigh about a pound, and large insects like dragonflies and grasshoppers weigh a fraction of what a mouse weighs. Small animals have certain advantages over large ones. In fact, the success of insects has a great deal to do with their diminutive size.

Perhaps the most important advantage of being small is that little animals, like insects, possess enormous strength relative to their weight. This is because an animal's weight increases in proportion to the cube of its size (by a factor of 3), whereas an animal's strength is related to the surface area of its muscles, which increases in proportion to the square of its size (by a factor of 2). Thus, the smaller an animal is, the stronger it is relative to its weight. Because of this, it is relatively easy for insects to carry objects that are many times their weight, or to jump many times their length.

Being small, insects can react much faster than larger animals. This is because reaction time is dependent on the speed of a nerve impulse and how far the impulse travels. Because insects are so small, their nerves are very short and so relatively little time elapses between when they perceive danger and when they react to it. This often gives an insect the advantage over a larger animal, such as a bird, that is trying to catch it.

Communication

We are primarily dependent on our vision, hearing, and speech to understand our surroundings and communicate with one another. On the other hand, many insects depend primarily on their sense of smell to perceive their surroundings and communicate. In insects that are highly dependent on their sense of smell, most of their brain may be devoted to interpreting signals from their olfactory (smell) receptors.

To communicate with others of their species, insects secrete various volatile hormones called pheromones. To attract a mate, female insects of many species secrete "sex pheromones." Males of their species can detect minute quantities of pheromones in the air and follow the scent to females. When threatened, many insects secrete "alarm pheromones" to warn others of their species of the danger. Many insects that congregate secrete "aggregation pheromones" that attract others of their species. There are also pheromones that insects employ to mark their territory and pheromones that some insects secrete to mark their path for members of their colony to follow. Queen bees and queen wasps of some colonial species secrete a pheromone that inhibits workers from constructing a new queen cell so that there is only one queen.

Pheromones tend to be small molecules because they diffuse faster than large molecules. Generally, each species secretes different concentrations of the same, or slightly different, pheromones so that the signal will be recognized only by members of their species. In addition to using their sense of smell to communicate with one another, many insect species use smell to identify plants or animals on which they feed or lay their eggs in or on.

Sensing Surroundings

The exoskeleton of most adult insects is covered with minute hairs called setae, each of which is connected to a sensory nerve. In addition to setae, insects have numerous other small sensory organs. Many setae and other sensory organs are concentrated on antennae, the insect's primary sensory organ, but they also are situated on the insect's legs, on sensory structures near the mouth of most insects called palpi, at the end of their abdomen called cerci, and on tubular organs called ovipositors through which female insects deposit their eggs.

Setae and other sensory structures can be found virtually everywhere on the insect's body. Those on the legs detect vibration that may be caused by a nearby predator or inform the insect of its position so it knows where to place its feet. Many of the setae and sensory organs on the antennae of insects measure minute amounts of various volatile chemicals emitted by other insects of its species to communicate or volatile molecules produced by plants or animals. Some sensory organs detect water vapor, as most insects need to find water. There are also setae or other sensory structures that measure temperature and wind direction and velocity. In some insects, setae or other sense organs detect sound.

Insect eyes are referred to as compound eyes because they are composed of a few to thousands of individual eyes called ommatidia. Each ommatidium produces an image with color and intensity such that an insect sees a mosaic composed of the images of individual ommatidia. You might think of the individual images as pixels. Generally, the more ommatidia an insect has, the better it can see.

Insects that are especially dependent on sight have large eyes that often cover most of their head, and the majority of their brain is often devoted to interpreting signals from their eyes. The eyes of insects are dome-shaped because each ommatidium is pointed in a slightly different direction than the adjacent ones. Because of this, many kinds of insects can see in virtually all directions without moving their head. However, even insects that have the best vision cannot see anywhere near the detail that we can. Also, insect eyes cannot accommodate to focus on objects at different distances.

In addition to compound eyes, many insects have two or three simple eyes on the top of their head called ocelli. The function of these simple eyes has been debated for many years. It has been proposed that they may function in detecting movement or helping flying insects to navigate.

Insects detect a somewhat different range of wavelengths than we do. Many insects cannot see red, but they can detect different wavelengths of ultraviolet light. Thus, an insect may be able to distinguish between flowers that appear the same color to our eyes because the flowers reflect different wavelengths of ultraviolet light. Many insects can also detect the plane of polarized light. On a sunny day, this enables flying insects to determine the orientation of their body with respect to the earth's surface, which helps them to navigate.

Insect eyes are particularly good at detecting movement, which helps them to perceive an approaching predator, or prey if they are predatory insects. Since their eyes are focused on objects that are just a few inches away or less, they cannot visualize an object that is far away, such as an approaching person with an insect net. However, they can detect that something large is moving, and that's enough to cause them to hide or fly away.

In many insect species, males, and sometimes females, produce sounds to attract mates. Most of these sounds are made by rubbing various parts together (stridulation), although some insects produce sound by rubbing a part against a tree, plant, or the ground. Although we can hear the calls that are made by some insects like crickets and grasshoppers, the sound that is produced by most insects is not audible to our ears because it is too high a frequency or not loud enough. In some insects that produce sound, females need accurate directional hearing in order to locate a male that is producing the sound. Although we are usually able to discern the general direction that sound is coming from, it is only approximate, as anyone knows who has tried to find a bird in a tree from its call. Some insects have far more precise directional hearing than we do. Some species are able to fly in almost a direct line to find a mate or a prey animal.

Some of the insect's setae and small sensory organs are used to feel their surroundings and detect movement, especially vibrations that are caused by other animals. The ability to feel their way is essential to many nocturnal insects and insects like ants and termites, as there is no light in their underground nests. We are not so used to feeling small vibrations, but when they walk, even very small insects cause faint vibrations that can be sensed by nearby insects or certain other animals. Spiders are particularly good at detecting vibrations, which is essential because most kinds of spiders have very poor vision. Detecting vibrations is how web-building spiders detect insects that fly into their web and how spiders that hunt on the ground detect approaching insects.

Reproduction

For most insects reproducing requires both luck and skill. Consider a male insect searching for a mate in a field where there are only a few females of his species. Because most adult insects live for only a few days or weeks, the male doesn't have much time to locate a female, even if he is lucky enough to elude the many predators that live in the field. After mating, a female usually must locate a suitable place to lay her eggs. Many insects need to attach their eggs to one species of plant or lay their eggs in or on a particular kind of insect or another animal.

Rather than having an ovulatory cycle like most mammals, female insects that have mated are fertile 24-7 because after mating, sperm are stored in receptacles, and fertilization takes place when eggs are laid. Since most insects are short-lived, females do not need to store sperm very long, but female insects that live a long time as adults can store sperm for weeks or months. Thus, although it may be difficult for male and female insects to find each other, once a female has mated, she can lay fertile eggs for the rest of her life.

PHOTOGRAPHING INSECTS

As many professional and amateur photographers photograph insects, there are a number of websites and blogs devoted to insect pictures. Photographers employ a variety of equipment and techniques. I will explain how I have photographed the insects in this book, but I suggest that the reader who is interested in taking pictures of insects look at some of the different techniques on YouTube and various blogs and websites, as well as books and articles on macrophotography.

One difficulty with photographing insects is that they are usually very active. A small insect can move out of the field of view in an instant. Of course, an insect can also fly or run away and disappear before you can take a picture. Thus, one can usually not use a tripod or set up lights to obtain the best lighting. I prefer to photograph on relatively cool days because being cold-blooded, insects are less active on cool days. Wind can also be a problem. Even a small breeze that moves a plant or leaf back and forth a fraction of an inch can make photography of a tiny insect difficult and frustrating. I usually do not try to photograph insects on windy days.

Another problem with photographing tiny objects is that the smaller an object is, the smaller the depth of focus. Thus, when photographing insects, I usually stop down the lens to a small aperture. Also, it is important to take the picture from the best angle. Just as with portrait photography, it is usually essential to focus on the insect eyes, because when we look at a person or an animal, we focus on the eyes. Even with the smallest aperture, the entire insect is usually not in focus. Thus, if you photograph an insect with its head facing you, most of the insect will be out of focus, so it is usually best to take the picture from the side or from above. With experience, one learns how the picture will look from what he or she sees in the viewfinder or screen.

I photograph insects with a (Nikon D610) full-frame digital single-lens reflex camera using a (Nikon 105mm or Sigma 150mm) macro lens. Macro lenses are designed for photographing small objects. I use a relatively long focal length lens to take pictures of these small animals because there is more "working distance" between the lens and the insect. With a shorter focal length lens, the front of the lens may be so close to the insect that it hits the surface the insect is on or nearby leaves before the insect is in focus. In addition, the farther the lens is from an insect, the less likely it is to frighten it away. For the smaller insects, I place (Kenko) extension tubes between my camera and the lens to facilitate filling the field with these tiny animals. With the smallest subjects, I may need several extension tubes.

As there is very little depth of focus with tiny objects, I usually stop down the 105mm lens to a small (f/22, f/25, or f/32) aperture in order to get most of the insect in focus. (As a rule of thumb, the smaller the insect, the more I stop down the lens.) There is also very little light with small objects, especially when using small apertures, so it is usually necessary to use a flash. The (Sigma EM-140) ring flash, which is designed for photographing small objects that are very close to the camera lens, works well for me.

Some of the pictures in this book were taken in the field, while others were taken at home. Some aspects of insect life must be captured where the insects live; for example, when insects are feeding or mating. If I were to move these insects, they would most likely stop what they are doing. However, I usually catch small insects by sweeping a net through plants and bushes and then transfer the insects to small (Darice Clear Bead Container) plastic containers. These insects are brought back home to be photographed.

Many insects are very difficult to see in nature because of their size or their camouflage. Others fly away from my presence. Since insects usually run or fly away when I open their container at home, I have found that putting the containers in a refrigerator overnight at 40 degrees Fahrenheit slows them down long enough for me to place them on a leaf or stick and take their picture. However, some insects fly away as soon as they warm up. Sometimes I am able to recapture them and put them back in the refrigerator for a longer time, but often they escape. My wife sometimes finds them, but more often it's the cat, who loves to catch them for a tasty treat.

When a day of photographing is done, I release the insects in a little wooded area behind our house. I could return each insect to the spot where I found it, but that's a bit much.

NOMENCLATURE

All insects that have been described have a Latin name (genus and species); more familiar insects also have a common name. Most of the 90,000 North American insects that have been described are relatively rare and so have no common name. (There are also estimated to be many thousands of North American insects that have not yet been described.) Scientists use the scientific (Latin) name when they write or talk about insects, and since they are uncommon, these insects never got a common name. With the exception of dragonflies and butterflies, common names are not standardized and can be different in different regions of North America. Also, many insects have more than one common name, and in some cases, two insects may have the same common name. Most of the insects in this book are relatively common, so they have a common name.

A professor of mine talked about entomologists as either lumpers or splitters. Lumpers are those that tend to consider very closely related insects as a single species, whereas splitters are those that consider them to be separate species. DNA sequencing has provided a new tool for determining relationships between insects. Some insect systematics put a lot of faith in DNA sequencing, but others rely more on anatomical features. Since evolution is a continuous process and there is no precise definition of what constitutes a class, order, family, genus, or species, there is considerable controversy. To make matters even more confusing, insects are frequently reclassified such that a group of insects that were considered separate genera or species become reclassified as a single genus or species. In a similar way, several insects that were considered the same genus or species become reclassified as different genera or species.

It is easy to determine the species of some insects. For example, many people can identify monarch butterflies (*Danaus plexippus*) because of their distinctive wing pattern. However, many insects look very similar to closely related species. Most genera contain a number of species, and in most cases, only an expert in the particular group of insects can tell them apart. To do this, he or she kills the insect, often dissects it, examines it very carefully under a microscope, and compares it to descriptions and drawings in very specialized texts. Thus, in cases when I am not sure of the species, I only identify it to genus and write "sp." after the name of the genus, meaning that I cannot determine the species.

1
SPRINGTAILS

Springtails are minute, wingless, six-legged arthropods. Most species live in soil or leaf litter, where they feed on bacteria, fungi, lichens, and decaying vegetation. About 700 species of springtails have been described in North America.

SLENDER SPRINGTAILS

Latin name: *Pogonognathellus* sp.

Family: Entomobryidae

Identification: 0.03 to 0.2 inch long, dark-colored, long antennae

Distribution: Throughout North America

Comments: These tiny arthropods used to be considered insects, and you will find them in older books about insects. However, most experts now agree that they are a separate group of six-legged arthropods. Experts frequently move animals into different groups based on new evidence. To make matters more confusing, experts often disagree. Maybe one day springtails will be reclassified as insects.

Springtails get their name from a structure under their body called a furcula that they can use to spring into the air. Although springtails are extremely common, you are not likely to see one because most species are minute and blend in with the soil where they live. However, they are relatively easy to find by scooping up some soil or leaf litter, placing it on a large piece of white cardboard or paper, and using a small paintbrush to move the soil or leaf litter around. If you don't want to sift through soil, you can purchase them on Amazon or other websites. Hobbyists use them to feed their poison dart frogs.

2
SILVERFISH

Silverfish are small, wingless, scaled insects with three long tails. They have flat bodies and flexible exoskeletons that enable them to fit into tight spots where predators cannot reach them. Unlike most insects, silverfish molt throughout their lives. Outside they live in damp logs or mossy places. They are called silverfish because they have a shiny appearance, caused by light reflecting off of their scales, and their movement, which is a bit reminiscent of a fish because when they walk, their abdomen moves back and forth. Although they don't bite or carry disease organisms, silverfish are still a problem because they can feed on carpets, clothing, coffee, glue, paper, and photographs.

SILVERFISH, FISHMOTH

Latin name: *Lepisma saccharina*

Family: Lepismatidae

Identification: 0.3 to 0.5 inch long, shiny gray or light brown, wingless, tapering abdomen, 3 long tails

Distribution: Throughout North America

Comments: Silverfish are primitive insects that are believed to have been on Earth for 400 million years. Their appearance and habits can tell us some of what was going on in the insect world when the first insects evolved. For example, the way that silverfish reproduce is different from that of most other insects. When a couple meets, they caress each other with their antennae. The female then runs away, and the male chases and catches up with her. The male then lays a capsule containing sperm on the ground, which the female picks up with her genital opening. Although this way of reproducing is unusual for an insect, scorpions and their close relatives, as well as some species of spiders, also place a sperm capsule on the ground, which their mates pick up with their genital openings. Thus, this may be the way that primitive insects and their relatives transferred sperm.

3
EARWIGS

Some references state that earwigs get their name because of an old wives' tale that they climb into people's ears when they are sleeping, but other sources say that the name was derived from the shape of their wings and was later corrupted from "ear winged" to "earwig." These nocturnal insects hide during the day under rocks, flowerpots, boxes, and other dark places. They are often seen scurrying away when their hiding place is moved. They are easily identified by the prominent forcepslike appendages on the rear of their abdomen called cerci, which are larger and more curved in males. Although cerci are usually sensory structures, earwigs employ their cerci for defense and to unfold their hindwings, although they seldom fly.

Some earwig species are among the relatively few nonsocial insects that exhibit maternal care. Pairs of earwigs dig a nest in the ground. After the female lays her eggs, she chases her mate from the nest and remains with the eggs. She may leave the nest to feed but soon returns to guard her brood. When the eggs hatch, she continues to clean and care for her young until they undergo their second molt and fledge.

European earwig, male. Note large cerci.

EUROPEAN EARWIG

Latin name: *Forficula auricularia*

Family: Forficulidae

Identification: 0.5 to 0.6 inch long, brown, light brown legs, males with prominent cerci

Distribution: Throughout most of North America

Comments: This species was introduced into North America in the early twentieth century and has spread throughout most of the continent. Like other species of earwigs, European earwigs feed on small insects. However, when small insects are not readily available, they feed on plants and can become a pest in gardens and on farms, where they damage vegetables including cauliflower, celery, potatoes, beets, and cucumbers. They also may feed on corn and fruits, especially apples and pears, and they sometimes chew flowers, including roses and zinnias. However, because earwigs are usually predacious, they have been used to control aphids and other small crop pests. They usually do not harm plants when aphids or other small soft-bodied insects are plentiful.

4
TERMITES

Termites are colonial insects. Their colonies are composed of different castes, the most abundant of which is the worker caste, whose members are usually blind. They maintain the nest. The soldier caste, which is also typically blind, protects the nest and the queen and king who produce all of the young. Soldiers have large, hard heads and very large mandibles. They cannot feed themselves, so they are fed by workers. There are also winged reproductives that are capable of establishing new colonies. Unlike colonial bees, wasps, and ants, where workers are all sterile females, termite workers and soldiers are both males and females.

Termites evolved from cockroaches, which are their closest relatives. They feed on wood (cellulose), which is digested by protozoa (single-celled animals) and bacteria in their digestive tract since, like most other animals, termites cannot digest cellulose. Young termites obtain protozoa and bacteria by eating the feces of older termites.

To form new colonies, reproductive males and females fly from the colony and mate, often with reproductives from other colonies. After they mate, their wings fall off, and the pair walks off together in search of a suitable place to start their own colony. The large majority of these reproductives are eaten along the way by various predators, but a few lucky ones survive to establish a new colony. The first eggs that are laid by the female develop into workers that feed and care for their parents. Later, eggs become soldiers and the male and female that founded the nest become the king and queen of their new colony. Fed and cared for by the workers, the queen gradually grows to many times her original size. In fact, she becomes so large that she can no longer move. The queen usually lives for many years and may lay thousands of eggs a day. The faithful king remains by her side, cared for and fed by workers. He occasionally mates with the queen.

Most North American species of termites live in the southern United States. They are not native but were imported many years ago in lumber and wood that was used in shipping.

Reticulitermes flavipes, workers and alates

Reticulitermes flavipes, three workers and a soldier

EASTERN SUBTERRANEAN TERMITE

Latin name: *Reticulitermes flavipes*

Family: Rhinotermitidae

Identification: Workers 0.1 inch long, soft-bodied, creamy or gray-white round head, blind; soldiers 0.15 inch long, resemble workers but with larger, darker head and large black mandibles; reproductives (alates) 0.15 inch long, have eyes and wing-buds or wings

Distribution: Eastern United States, west to Texas, north to Ontario and Quebec

Comments: Subterranean termites cause many millions of dollars of structural damage to buildings every year. They enter buildings by tunneling through wood that is in the ground. There are now building codes that require homes and other structures to have cement foundations that are far enough off the ground that termites cannot tunnel into the home. However, sometimes dirt or wood can get piled up against the home or people make modifications that give termites an entry site. It is a good idea to look around the foundation of your home to see that there is no place that subterranean termites could enter. Termite infestation is often discovered by the presence of winged reproductive swarmers (alates) that have left their nest. Although termites can damage buildings, they also live in fallen logs, which aids in recycling trees.

5
LICE

Lice are tiny external parasites that live and feed on birds and mammals. Some feed on skin, fur, or feathers, while others, like those that live on humans, feed on blood. Almost all mammalian and avian species are infected by some kind of louse. Human body lice are uncommon in North America since most people regularly wash themselves and their clothes, but they had caused havoc in the past, especially during World War I when soldiers lived in trenches for long periods of time and were often unable to bathe or wash their clothes. Trench fever, which is caused by a bacteria, *Bartonella quintana*, which is transmitted by body lice, was rampant among soldiers. Although the disease was rarely fatal, infected soldiers were usually unable to fight effectively. Typhus, also transmitted by body lice, was somewhat less common, but it is a far more serious and often deadly disease. Thus, body lice were probably responsible for battles being won and lost. Body lice were not much of a problem for most of World War II because they were killed by DDT, which was invented in 1939.

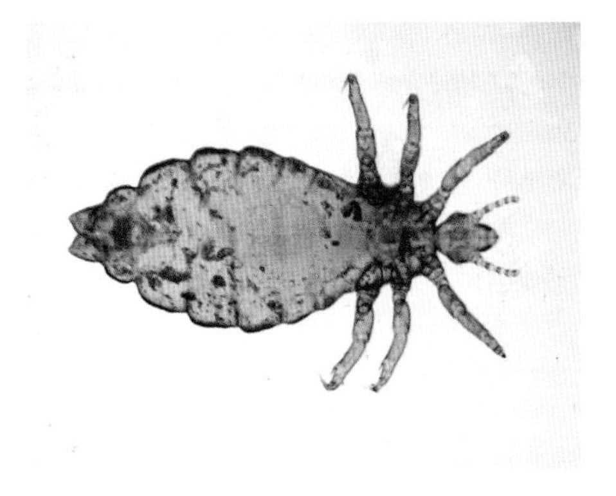

CRAB LICE, PUBIC LICE

Latin name: *Phthirus pubis*

Family: Pediculidae

Identification: 0.04 to 0.08 inch long, round body, front legs thinner than other legs

Distribution: Live exclusively on humans worldwide

Comments: Crab lice live on humans, where they feed on blood. They are obligate external parasites that will die in a day or two if they are removed from the body. Crab lice are usually found on pubic hair, although they are sometimes found on eyelashes and on areas with coarse hairs such as beards, mustaches, and underarms. However, they are not found on finer hair such as hair on the scalp. Crab lice are usually spread by sexual contact, although they can they can be spread by bedding, clothing, and shared towels. The main symptom of an infection is itching that is caused by a reaction to the saliva of the lice. However, crab lice are not known to transmit diseases. The presence of these lice can be determined by examining hairs for their eggs, called nits, which look like minute pussy willows. The infestation can be treated and usually eliminated with a number of products that can be purchased without a prescription.

HUMAN LICE

Latin name: *Pediculus humanus capitis, Pediculus humanus humanus*

Family: Pediculidae

Identification: 0.08 to 0.15 inch long, tan or gray-white, large oval-shaped abdomen

Distribution: Head lice common in North America, body lice uncommon

Comments: Head lice (*Pediculus humanus capitis*) and body lice (*Pediculus humanus humanus* or *Pediculus humanus corporis*) are closely related and look the same. They both feed on blood. Head lice infect several million people each year in North America. Infections cause itching and can result in loss of sleep, but they are usually not serious, and head lice do not transmit any disease in North America. Head lice are spread by direct contact or sharing towels and bedding. They are especially common in children, who acquire them from their friends and schoolmates. Treatments can be purchased without a prescription. Body lice are uncommon in North America.

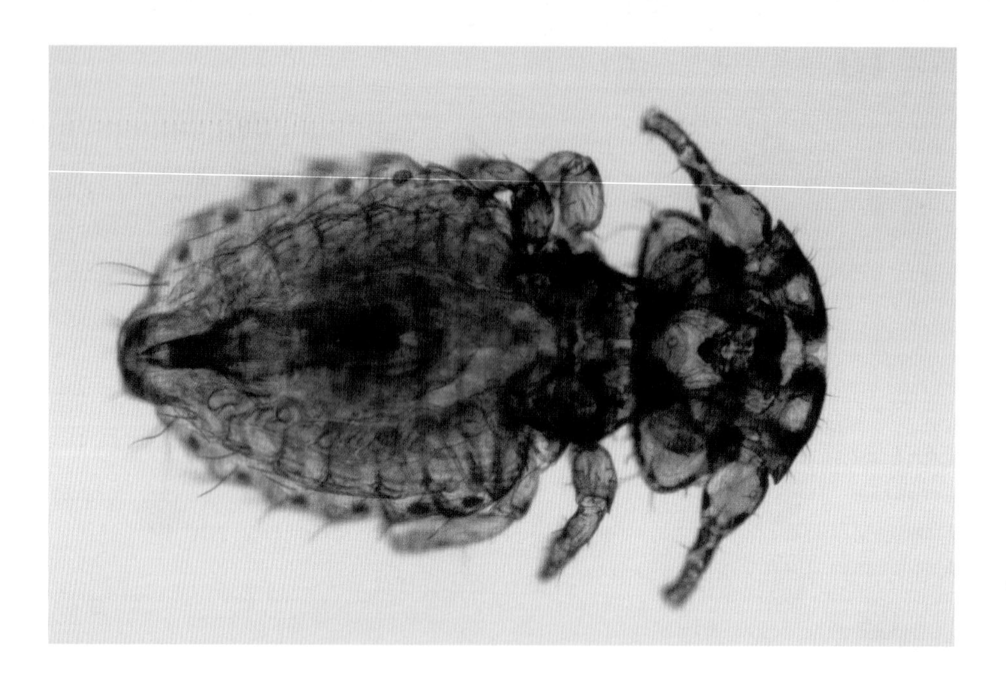

DOG LICE, CANINE CHEWING LICE

Latin name: *Trichodectes canis*

Family: Trichodectidae

Identification: 0.06 inch long, yellow, head broader than long, legs stout with single claw, thick setae

Distribution: Live exclusively on dogs and other canines worldwide

Comments: In North America, dog lice are found on canines, including dogs, wolves, and coyotes. They live their entire life on the host but can be transferred to uninfected dogs by close contact or shared combs or brushes. Although they do not usually cause major problems on healthy dogs unless the dog is heavily infested, they can have a serious effect on older dogs, puppies, and debilitated dogs. They can also infect canines with the dog tapeworm *Dipylidium caninum*.

Female lice usually lay several eggs a day for about a month. The eggs take a week or two to hatch into nymphs that mature to adults in about two weeks. Infected dogs exhibit excessive scratching and biting. Lice and their eggs can be found by examining hairs with a microscope or strong magnifying glass. The infestation can be eliminated by treating the dog with one of a number of insecticides, including permethrin, carbaryl, and fipronil.

6
FLEAS

Fleas are small, flightless, external parasites of mammals and birds. The bodies of fleas are flattened laterally, which allows them to fit between hairs or feathers. Prominent claws enable them to hold on to their host. They feed on blood. Most species live on a specific host, but some parasitize a number of different hosts.

RAT FLEA, ORIENTAL RAT FLEA, TROPICAL RAT FLEA

Latin name: *Xenopsylla cheopis*

Family: Pulicidae

Identification: 0.1 to 0.2 inch long, flattened side to side, wingless, lacks the combs that are present in other species of fleas

Distribution: Throughout North America

Comments: The Oriental rat flea is the primary vector of *Yersinia pestis*, the bacterium that caused bubonic plague. Rats are immune to bubonic plague but carry the bacterium. The disease occurs when a rat that carries the bacterium bites a human. Bubonic plague can also be transmitted directly from one person to another, and it is not known how many infections were caused by the rat flea. It is estimated that the bubonic plague killed 350 to 450 million people in the fourteenth century and that it took 200 years for the world population to recover.

This rat flea can also transmit murine typhus (also called endemic typhus), which is caused by a bacteria, *Rickettsia typhi*. Most people who are bitten by rat fleas do not realize that they have been bitten. Their symptoms are similar to other diseases. Fortunately, if the infection is diagnosed, it can be easily cured with antibiotics.

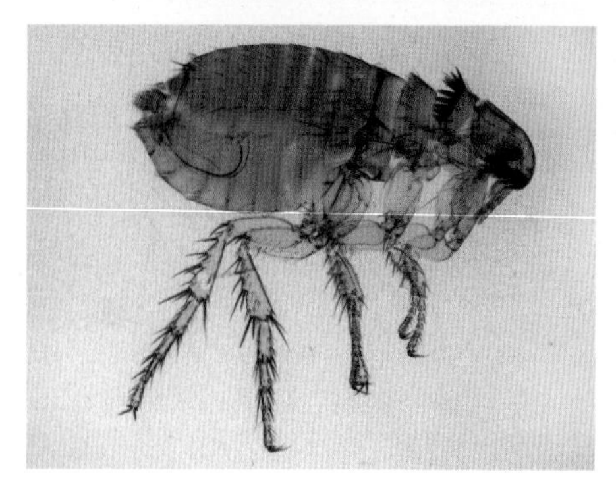

CAT FLEA

Latin name: *Ctenocephalides felis*

Family: Pulicidae

Identification: 0.07 to 0.1 inch long, dark-colored, body compressed laterally; unlike other fleas, cat fleas have only simple eyes

Distribution: Wherever there are cats or dogs

Comments: Although they are called cat fleas, *Ctenocephalides felis* can infect dogs. However, the fleas that usually live on dogs are dog fleas, *Ctenocephalides canis*. The bodies of these and other fleas are covered with many spines that are directed backward, which helps them to stay on their host and hold on when they are feeding. They have a tough, rubberlike exoskeleton containing the elastic protein resilin that protects them from being injured when their host scratches. They also make use of resilin when they feed, as all fleas feed on blood. When they are hungry, they lift their abdomen up so their head is pressed against their host. Inside their head is a structure that is something like a slingshot, and the flexible part that powers the slingshot is made of resilin. The rapid contraction of resilin forces a knifelike structure into the host. The fleas then lap up the blood from the wound.

7
BARKLICE AND BOOKLICE

Barklice and booklice are small soft-bodied insects with chewing mouthparts. There is often considerable variation in the shape of individuals of the same species. They can be winged or wingless. Winged forms hold their wings rooflike over their bodies.

BARKLICE

Latin name: *Psocus* sp.

Order: Psocoptera

Identification: 0.04 to 0.2 inch long, large eyes, threadlike antennae, wings rooflike over abdomen

Distribution: Throughout North America

Comments: Barklice are not lice, and they are not related to lice, although there is evidence that lice may have evolved from barklice. They are very primitive insects that first appeared in the Permian period, 295 to 248 million years ago. Although a few species live in buildings and can be pests, most species of barklice live in groups on the base of trees, where they feed on fungi, algae, and lichens. Groups of barklice often spin silk webs for protection from predators and to prevent desiccation. They flee from danger by leaping backward or flying.

BOOKLICE

Latin name: *Psocus* sp.

Order: Psocoptera

Identification: 0.04 to 0.2 inch long, threadlike antennae, wings absent or greatly reduced

Distribution: Throughout North America

Comments: Booklice are closely related to barklice. They are so named because they inhabit public and private libraries, where they feed on the glue in book bindings. They can spread from one library to another and to homes when books are checked out and taken home, or read in parks or other places where booklice live and feed on lichens and fungi. However, they require a fairly humid environment, so air-conditioning has greatly reduced or eliminated them from many libraries and homes. Booklice are also common in new homes because new homes can be relatively humid, and booklice can inhabit wood that is used for construction. They can be eliminated by removing mold or fungus and turning up the heat, which will reduce the humidity. Using a dehumidifier will also kill booklice.

8
PSYLLIDS

Psyllids are small, plant-feeding insects. Each species feeds on only one species of plant or a few related species. Some authors consider them to be primitive hemipterans.

PSYLLIDS, JUMPING PLANT LICE, PSOCIDS

Latin name: *Psyllidae* sp.

Family: Psyllidae

Identification: 0.08 to 0.2 inch long, resemble miniature cicadas

Distribution: Throughout North America

Comments: Psyllids are tiny winged insects that jump or fly away when disturbed. Nymphs are flat and do not look like adults until their final molt. Adults and nymphs suck plant juices. Each species feeds on only one species of plant or several related species. Although the plants that most psocids feed on are of no economic importance, some species cause a great deal of damage because they feed on fruit trees or potatoes. Some species secrete a wax that covers their surface for protection.

9
THRIPS

Thrips are tiny, slender, cigar-shaped insects with fringed wings. They can be extremely numerous on flowers and in fields, where they feed on plant juices. They have chemicals on their feet that can irritate your skin if many of them land on you.

Thrips on a daisy

THRIPS

Family: Thripidae

Order: Thysanoptera

Identification: 0.02 to 0.05 inch long, thin, cigar-shaped, fringed wings

Distribution: Throughout North America

Comments: You may have never noticed a thrips because they are tiny. (Thrips is both singular and plural.) However, thrips are some of the most destructive pests of crops and fruit trees. They have asymmetrical mouthparts: Their right mandible is reduced and nonfunctional, while the left mandible is used to cut the plant tissue to release the sap on which they feed. Thrips also have strange feet with a bladderlike structure that can be inverted, enabling them to walk on vertical surfaces. Thrips wound the plants that they feed on and, in the process, can spread a number of plant pathogens. However, not all species of thrips are pests. Some kinds are beneficial because they pollinate plants or feed on fungi or small insects, including other species of thrips.

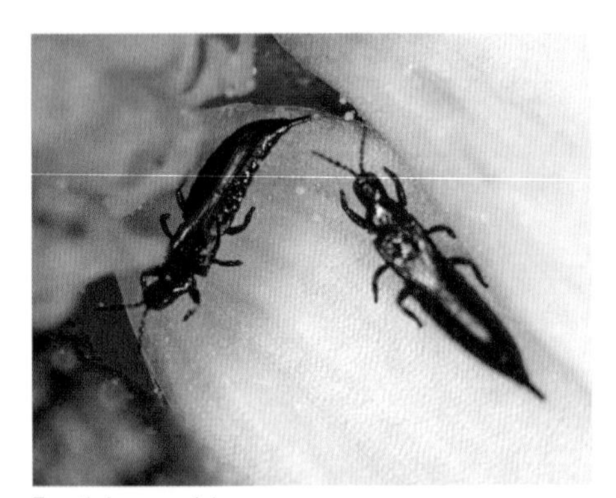

Two thrips on a daisy

10 SCORPIONFLIES

There are fifty-four species of scorpionflies in North America. They are only found in eastern North America, and all are in the genus *Panorpa.* Scorpionfly larvae resemble the caterpillars of moths and butterflies; however, unlike the caterpillars of moths and butterflies that have simple eyes, scorpionfly larvae have compound eyes.

Female scorpionfly

SCORPIONFLIES

Latin name: *Panorpa* sp.

Family: Panorpidae

Identification: 0.4 inch long, orange, black markings on wings, long proboscis

Distribution: Eastern North America

Comments: Scorpionflies are named for the male genitalia, which looks like the stinger on the end of the long abdomen of scorpions. Although it is reminiscent of a stinger, the structure is a copulatory organ. Scorpionflies are harmless. They feed on dead insects, sometimes stealing them from spiderwebs. Scorpionfly larvae resemble caterpillars but can be distinguished by their eight pairs of prolegs (caterpillars have a maximum of five pairs). Males of many kinds of animals present females with a "nuptial" gift to gain their favor. With some species of scorpionflies and certain other insects, the gift is a dead insect. If the female likes the gift, she consumes it while mating, but if the dead insect doesn't look that tasty, she flies off. Consuming the nuptial gift induces egg production.

Male scorpionfly

11 DRAGONFLIES

Dragonflies are predacious both as larvae and adults. They are considered primitive insects because they have been around a long time. Flying insects that have four wings use their wings differently. For example, beetles use their forewings for lift and steering, like the wings of an airplane, and their hindwings for propulsion. Other insects, like moths, have hooks or other means of attaching the two wings on one side together so they beat as one. Dragonflies have the ability to move each of their four wings independently. This gives them the ability to turn on a dime, fly backward, and hover so that they can outmaneuver birds and catch the flying insects that they feed on. However, dragonflies cannot walk or run, so they do all their activities in flight or when they are perched.

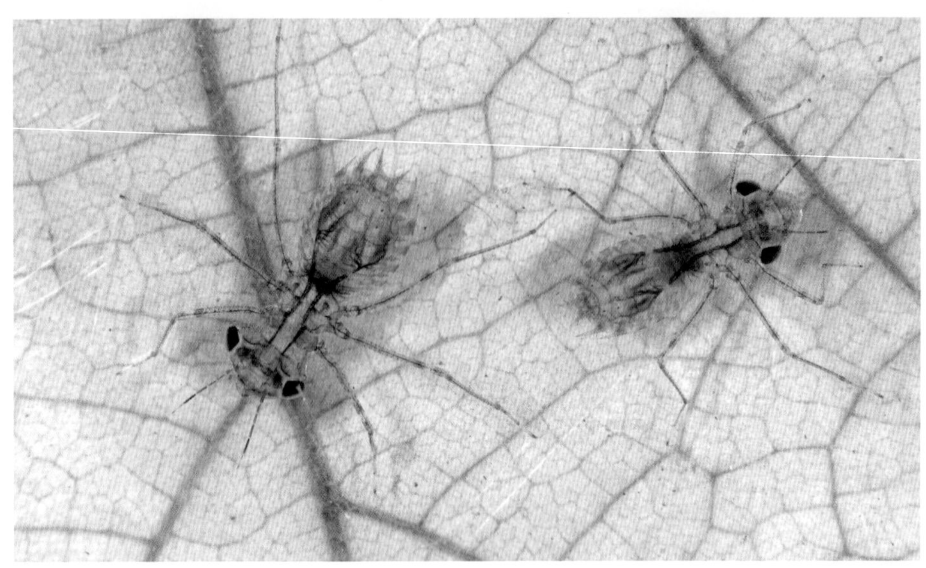

Two young naiads

DRAGONFLY NAIADS

Latin name: *Odonata* sp.

Family: Libellulidae

Order: Odonata

Identification: Length depends on instar and species, green or brown, wide head, large eyes

Distribution: Throughout North America

Comments: In Greek mythology, a nymph is a spirit in the guise of an attractive maiden that lives in forests or mountains. Naiads are nymphs that inhabit lakes, streams, and rivers. The larvae of insects that undergo simple metamorphosis are usually termed "nymphs." However, if they are aquatic, they are usually referred to as naiads, although some authors call them nymphs. Dragonfly naiads are aquatic. Unless you want to wait for it to become an adult or you are an expert, it is usually not possible to determine the genus or species of a dragonfly naiad. Dragonfly naiads are predacious, just like adults. They are mostly ambush hunters that wait for insects, small fish, or amphibians to pass by. When the prey is near enough, they rapidly extend their hinged lower lip (labium) and scoop up the prey.

Females of most species of dragonflies deposit their eggs in the water or on aquatic plants. When they hatch, the naiads undergo a dozen or more molts before their final adult molt. Depending on the species and temperature of the water, they may spend from a month to several years as naiads. Like most other naiads, dragonflies use gills to breathe oxygen that is dissolved in the water. In the case of dragonfly naiads, their gills are located in their hindgut. Expanding and contracting their hindgut circulates the water. Dragonfly naiads are also capable of expelling water rapidly to shoot them forward in order to escape danger. They often bury themselves in mud at the bottom of the ponds that they live in so they will not be detected by predators.

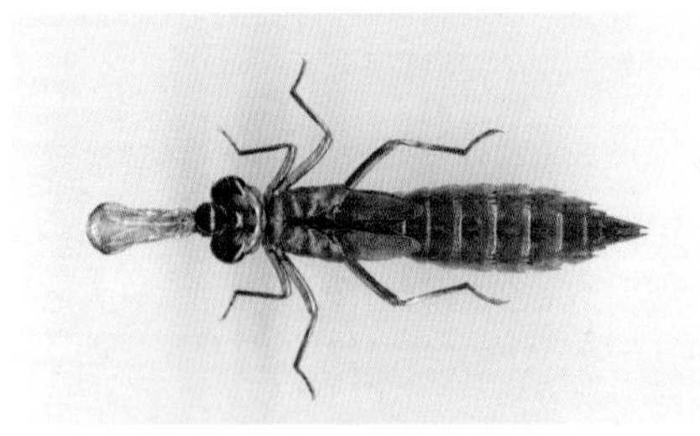

Naiad with jaw (labium) extended

After a naiad has undergone its last larval molt and is ready to become an adult, it stops feeding and climbs onto a plant, where it positions its legs firmly on the plant and molts into an adult dragonfly. However, although they are adults, it takes young dragonflies a few days to become fully mature and develop their adult colors.

EASTERN AMBERWING

Latin name: *Perithemis tenera*

Family: Libellulidae

Identification: 1 inch long, orange wings and legs, orange abdomen with brown, yellow, and white pattern

Distribution: Eastern United States to Texas, Colorado, and Wyoming; throughout Ontario

Comments: The eastern amberwing belongs to a family of dragonflies called skimmers, which tend to be very colorful. Like most skimmers, eastern amberwing dragonflies perch after short flights and usually return to the same perch. Most dragonflies are not mimics, but this dragonfly, with its brown and yellow pattern, is a wasp mimic. To make the disguise more convincing when they are perched, eastern amberwing dragonflies wiggle their abdomen and wings as a wasp might do. Both males and females usually hunt some distance from the water. Males search out a good place for a female to lay her eggs, usually a spot where there are lily pads or floating algae. They then choose a lookout spot and wait for a female. After mating, the male leads the female to the chosen spot.

COMMON WHITETAIL

Latin name: *Plathemis lydia*

Family: Libellulidae

Identification: 1.6 to 1.9 inches long, males have a white abdomen and a black area in the middle of their 4 wings, females are brown with small white spots on their abdomen and 3 black areas on their wings

Distribution: Throughout the United States and southern Canada

Comments: Unlike most dragonflies in the skimmer family, common whitetails rest on the ground, logs, or rocks in open areas. Males ward off rivals by flashing their white abdomen. After mating, the female deposits her eggs by tapping the water with her abdomen, often with her mate flying above her or nearby. Each time her abdomen touches the water, she lays 20 or so of her approximately 1,000 eggs. Larvae of common whitetails are relatively tolerant of polluted water.

BLUE CORPORAL

Latin name: *Ladona deplanata*

Family: Libellulidae

Identification: 1.3 to 1.5 inches long, flattened abdomen, blue on males, gray on females, brown with white stripe on immature females

Distribution: Southern Quebec to Florida, west to Texas, absent from Midwest

Comments: Blue corporals, like other skimmers, are often observed gliding over ponds and fields. Gliding is much more efficient than flying because it requires very little energy. In fact, some dragonfly species that live in the northern United States and southern Canada migrate south in the fall and spend the winter in southern states. They could not travel such long distances without the ability to glide efficiently.

Both dragonflies and damselflies have a dark, heavy spot called a pterostigma on the leading edge of each wing. It is most noticeable in clear-winged dragonflies like blue corporals. Certain other insects that glide like ant lions, owlflies, and snakeflies also have a pterostigma, but it is often not obvious. The pterostigma acts as a weight that prevents the wings from flapping when the dragonflies glide. This enables dragonflies to glide faster and more efficiently than they could without these structures.

Blue corporal eyes

SADDLEBAGS, SADDLEBAG GLIDERS

Latin name: *Tramea* sp.

Family: Libelluilidae

Identification: Saddlebags 1.6 to 1.8 inches long, saddlebag gliders 2 to 2.3 inches long, clear wings with black markings on the base of hindwings

Distribution: Throughout North America

Comments: Saddlebags are gliding dragonflies with dark markings at the base of their clear wings. When they fly, these markings create the illusion that they are carrying bags, which is why they are called saddlebags. They fly for long periods of time over fields and forests, even to the tops of trees. They usually perch on the tips of twigs with their abdomen dropped down. They lay their eggs in small ponds that do not have fish (for obvious reasons). Mating pairs usually perch for a few minutes. They then fly over the water, where the male releases his mate so she can tap the water and lay a few eggs before she returns to him.

12
DAMSELFLIES

Damselflies resemble dragonflies, but they are slimmer and smaller. They rest with their wings folded over their back rather than spread apart like dragonflies. Females of many species are difficult to identify because they have two or sometimes three different color forms.

Before mating, male damselflies transfer sperm from their penis, which is on the back of their abdomen, to a structure below the front of their abdomen termed the secondary penis. When courting, the male faces the female and puts his hindwings downward and his forewings and abdomen upward, revealing the lower part of his abdomen. The male then grasps the female behind her head with structures at the back of his abdomen and swings her abdomen around so that her genital opening attaches to his secondary penis. The male then inseminates his mate with sperm that he previously transferred to his secondary penis. The secondary penis also has a small brushlike organ that is employed to remove sperm from a previous mating to ensure that he will sire her offspring. Dragonflies also mate in this manner and have a brushlike organ on their secondary penis to remove sperm from a previous mating.

Damselfy naiad. Note three gills on abdomen.

DAMSELFLY NAIADS

Order: Odonata

Suborder: Zygoptera

Identification: Long and thin, with 3 feather-like gills extending from their abdomen. Length depends on instar and species

Distribution: Throughout North America

Comments: Damselfly naiads are slender, sensitive to oxygen levels, and, unlike dragonfly naiads, do not bury themselves in mud at the bottom of ponds, lakes, and rivers. These predatory naiads feed on mosquito larvae, daphnia (a small freshwater crustacean), and other tiny arthropods. When feeding, damselfly naiads extend their lower jaw (labium) rapidly to seize their prey. Damselfly naiads breathe through three gills that extend from their abdomen and are connected to their tracheal system. They also use their gills to propel themselves through the water as they wiggle their long abdomen.

When they are ready to molt, naiads climb out of the water and hold firmly onto a plant. Their thorax splits open, and the adult climbs out. Newly emerged adults pump hemolymph into their wings, which expands them. When their wings are fully expanded, they pump the hemolymph back into their body, which expands their abdomen. During the next few days, the exoskeleton hardens and the adult colors appear.

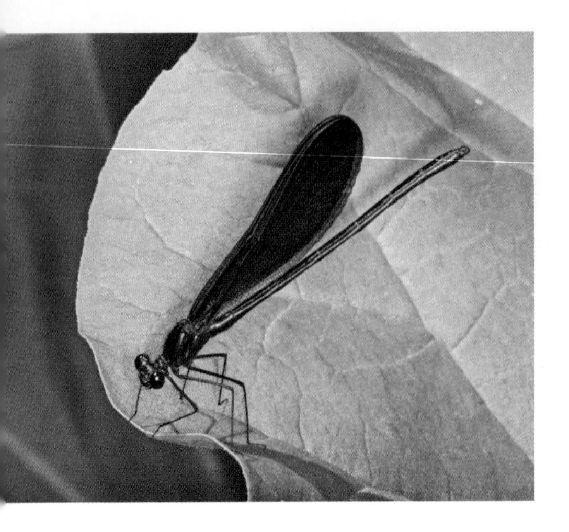

EBONY JEWELWING

Latin name: *Calopteryx maculata*

Family: Calopterygidae

Identification: 1.5- to 2-inch wingspan, black wings, blue body

Distribution: Northeastern United States and southern Canada, west to Texas

Comments: Ebony jewelwings are a species of broad-winged damselflies. They are easy to distinguish from other North American damselflies by their distinct dark wings. Males have green iridescent bodies, and depending on the light, they may also look blue or black. The wings are black because of their many black veins. The wings of the female are also black. Their bodies are bluish green, but they are not iridescent like males. These damselflies are territorial. Males fight off other males that come into their territory.

SPREADWING DAMSELFLIES, SPREADINGS

Latin name: *Lestes* sp.

Family: Lestidae

Identification: 1.6- to 1.8-inch wingspan when at rest, large pterostigmata

Distribution: Eastern North America and southeastern Canada

Comments: Most damselflies hold their wings together over their abdomen when they are at rest. However, spreadwing damselflies hold their wings at an angle away from their bodies when they are perched, although they hold them together at night, when it is raining, or when they are threatened. When they perch, they also hold their abdomen inclined down or vertically. The pterostigma of spreadwing damselflies is somewhat more prominent than that of other damselflies. Unlike most damselflies that overwinter as naiads, spreadwing damselflies overwinter as eggs, which are sometimes covered with snow. In the spring, the naiads hatch in meltwater pools, where they complete their development before the pools dry up.

ORANGE BLUET

Latin name: *Enallagma signatum*

Family: Coenagrionidae

Identification: 0.6- to 1-inch wingspan; male orange with black markings on thorax, black posterior of abdomen, orange legs; female pale yellow-green, blue, or orange

Distribution: Eastern North America, west to South Dakota and Texas

Comments: Orange bluets are active in the afternoon. Females usually stay some distance from the water unless they are mating or laying eggs. Like some other kinds of damselflies, females enter the water to lay their eggs, often accompanied by their mate. When they dive into the water, a layer of air around their bodies enables them to breathe through their spiracles. In fact, these damselflies can remain submerged for up to an hour. Although damselflies cannot walk, their diet is not limited to flying insects. Their superb flying ability allows them to hover next to plant stems while they snatch aphids from the plant.

CITRINE FORKTAIL

Latin name: *Ischnura hastata*

Family: Coenagrionidae

Identification: 0.8 to 1.3 inches long; male mostly orange, with an orange or brown tear-shaped forewing pterostigma, eyes yellow in front and green behind; female orange with broad, black mid-dorsal stripe and thin shoulder stripes, with an orange or brown tear-shaped forewing pterostigma

Distribution: Eastern United States, west to Texas and Nebraska

Comments: These small damselflies are called forktails because males have a forked projection at the tip of their abdomen. It is often difficult to identify females because their color is orangish when they are young but becomes darker as they age. Also, some females resemble males. The citrine forktail is the only North American species of damselflies with an orange or brown tear-shaped pterostigma on its forewings. Male forktails establish territories and choose breeding sites. If a male approaches a female and she does not want to mate, she curves the tip of her abdomen upward, signaling that she is not receptive. The male will then usually leave the female alone. After mating, females deposit their eggs one at a time on floating vegetation in ponds and lakes.

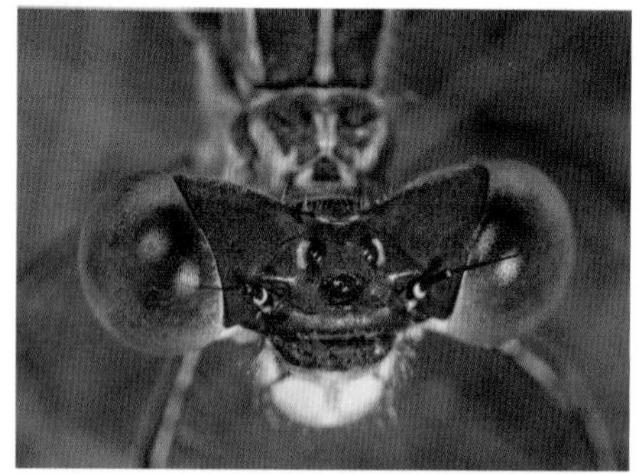

Damselfly head. Note two large eyes and three small ocelli.

13 MAYFLIES

Mayflies are short-lived insects. Aristotle is responsible for naming the order of mayflies Ephemeroptera, meaning short-lived. Indeed, adults of most species live only a day or two and do not feed. On the other hand, naiads may live for up to several years. Mayflies are not as common as they used to be, probably because mayfly naiads cannot survive in polluted water. Mayflies are the only insects that have two winged forms.

MAYFLY NAIADS

Order: Ephemeroptera

Identification: Aquatic, 2 or 3 long tails, gills on sides of abdomen, length depends on instar and species

Distribution: Throughout North America

Comments: The aquatic naiads of mayflies live in clear lakes, streams, and rivers, where they are an important food source for many freshwater fish. People who fish for trout often use flies that resemble mayfly larvae or adult mayflies. Naiads of most species graze on stones, where they scrape algae off of rocks and leaves with their mouth or their forelegs. However, they also feed on other plant matter and detritus. Naiads of some species also eat floating plants, and a few of the larger species prey on aquatic insects and crustaceans.

Mayfly naiads usually undergo from twelve to forty molts before they molt into terrestrial winged forms called subadults or subimagos. The subadults are sterile and live for up to a day before molting again into fertile adult mayflies.

CHOCOLATE DUN

Latin name: *Eurylophella prudentalis*

Family: Ephemerellidae

Identification: 0.2 to 0.6 inch long; 3 tails; males with large, spherical, red-brown eyes

Distribution: Northeastern United States, southeastern Canada

Comments: Like other mayflies, the hindwings of the chocolate dun are very small. *Eurylophella prudentalis* belong to the family Ephemerellidae, which are often referred to as spiny crawler mayflies. The naiad crawlers are found in fast-flowing streams and rocky-bottomed clear rivers, where they usually dwell in pools where the flow of water is relatively slow. If they are threatened, the naiads raise their tails in front of their bodies so as to look larger than they are and project their tails to poke the predator. The name "dun" is used by anglers to refer to subadults. After subadults molt into adults, anglers refer to them as "spinners."

PRONG-GILLED MAYFLY, EARLY BROWN SPINNER, BLACK QUILL

Latin name: *Leptophlebia cupida*

Family: Leptophlebiidae

Identification: 0.5 to 0.8 inch long, black with smoky-colored wings

Distribution: Northeastern and central United States, southeastern Canada

Comments: These large mayflies are found near lakes and streams, where their aquatic larvae live. The larvae are usually found clinging to rocks unless they are swimming. Larvae of the genus *Leptophlebia* can be recognized by the forked gills on their abdomen, which gave them the name prong-gilled mayfly. The mandibles of some species protrude to the side and are referred to as mandibular tusks.

SPECKLED DUNS

Latin name: *Callibaetis* sp.

Family: Baetidae

Identification: 0.3 to 0.5 inch long, speckled body, black markings on forewings, 2 tails

Distribution: Throughout the United States and southern Canada

Comments: *Callibaetis* naiads are found in slow-moving, weedy places in streams and rivers. After they undergo their second adult molt from subimago to fertile adult, male mayflies form swarms over rivers and streams waiting for females. Females fly into the swarm and find a partner. The long, thin tails of the couple act like parachutes to keep the pair aloft as they mate, and the female lays her eggs. Adult mayflies are also light because they do not feed, and their digestive tract is full of air. Mating and laying eggs does not take long, which is good because it must be completed before the pair falls into the water.

14 CADDISFLIES

There are about 1,400 species of caddisflies in North America. They are usually colored drab gray or brown and resemble moths.

The larvae of *Ochrotrichia* sp. construct their cases of sand grains, which they precisely glue together with their salivary gland secretions. Their cases are reminiscent of a stone wall constructed by a skilled stonemason.

The larvae of *Brachycentrus* sp. build their cases of minuscule twigs that they cut into precise lengths with their sharp mandibles. As the larvae grow, they affix increasingly larger twigs on top of their cases.

TUBE-MAKING CADDISFLY NAIADS

Latin name: *Limnephilidae* sp.

Family: Limnephilidae

Identification: Naiads dwell in tubes that they construct of silk, sticks, rocks, or snail shells. Length depends on instar and species

Distribution: Throughout North America

Comments: Many insects build cases and nests, but the little aquatic caterpillar-like naiads of tube-making caddisflies are some of the most amazing builders in the insect world. When the tiny first instar naiads emerge from the eggs, they begin to build a protective case in which they will spend their naiad life and pupate. Each species builds a characteristic case, although the structure of the case differs somewhat depending on the building materials that are available. The nests are made of sand grains, tiny sticks, pieces of leaves, or minute snail shells held together with saliva or silk. A few species build their cases entirely of silk. As the naiads grow, they enlarge their cases. Some cases are cone-shaped; others are cylindrical. However, whatever the material, these constructs are exceptionally precise. Although some caddisfly naiads feed on plant material, most are predators that have long raptor-like front legs that they employ to capture little insects or crustaceans that pass by.

Most caddisflies, like the genus *Agrypnia*, can be distinguished from moths because their wings are covered with hairs rather than scales.

CADDISFLIES

Order: Trichoptera

Identification: 0.5 to 1.3 inches long, mothlike, long slender antennae, wings with setae and without scales, wings held rooflike over abdomen, unusually drab colors

Distribution: Throughout North America

Comments: Caddisflies are related to moths and are often mistaken for them. Like most moths, they are nocturnal and are attracted to lights, which is where they are usually seen. However, the wings of caddisflies are covered with hairs rather than scales that characterize the wings of moths. Many species have long, hairy labial palpi near their mouths. Usually, adult caddisflies do not feed. It is difficult to distinguish different species; experts usually identify species by the structure of their mouthparts. Caddisflies have one generation a year and overwinter as naiads.

The wings of some caddisflies, like those of the genus *Hesperophylax*, have relatively few hairs.

15 COCKROACHES

Cockroaches are one of the least-liked insects because they invade homes, apartments, and other buildings. The species that are found in buildings are the American cockroach (*Periplaneta americana*), the German cockroach (*Blattella germanica*), and the Oriental cockroach (*Blatta orientalis*). Despite their names, all three species are believed to be native to North Africa and traveled from various places to North America and around the world many years ago as stowaways on sailing ships. However, less than 1 percent of cockroach species invade homes or could survive in them.

The cockroach species that live in homes and other buildings are nocturnal omnivorous insects that feed on everything that humans eat as well as many things that humans do not, such as garbage and pet food. They are not known to transmit any particular disease, but if they happen to feed in contaminated food, bits of the food can attach to them and be spread to food in the kitchen. Cockroach traps employ synthesized pheromones to attract cockroaches into little cardboard boxes that have sticky glues. The cockroaches stick to the glue and cannot escape. Other traps and sprays use insecticides to kill roaches.

Female Dusty cockroach with an egg case

DUSTY COCKROACH

Latin name: *Ectobius lapponicus*

Family: Ectobiidae

Identification: 0.2 to 0.4 inch long, brown flattened body, long antennae, 2 cerci

Distribution: Northeastern United States

Comments: Dusty cockroaches were first discovered in Maryland in 1985 and have since spread throughout the Northeast. Although they came to North America from Europe, like most other cockroaches, including those that are pests, the dusty cockroach is probably native to North Africa. Dusty cockroaches live in woodlands and don't enter buildings. Female cockroaches of this and a number of species carry their eggs in an egg case until they are ready to hatch. The disadvantage of this is that if a predator eats the female, the eggs will also be lost. The advantage of having an egg case over depositing the eggs is that a predator is less likely to find the eggs and eat them.

Cockroach feet have claws for gripping rough surfaces and pads for gripping smooth surfaces.

SPOTTED MEDITERRANEAN COCKROACH, WOOD ROACH

Latin name: *Ectobius pallidus*

Family: Ectobiidae

Identification: 0.2 to 0.4 inch long, light brown flattened body, long antennae, 2 prominent cerci; nymphs have many dark brown spots

Distribution: East Coast of the United States

Comments: Spotted Mediterranean cockroaches were first found in North America on Cape Cod, Massachusetts, in 1948 and have since extended their range. Like other species of cockroaches, spotted Mediterranean cockroaches are mostly nocturnal. However, it is safe for you to come to Cape Cod in the evening because these cockroaches live outside and do not come into buildings.

Have you ever tried to sneak up on a cockroach? It's not easy because cockroaches have sense organs on their legs, which detect the slightest movement on the ground or floor. They can also detect slight changes in air currents when someone approaches. Although *Ectobius pallidus* can fly on warm nights, like other cockroaches, they usually react to danger by scurrying for cover.

MADAGASCAR HISSING COCKROACH

Latin name: *Gromphadorhina portentosa*

Family: Blaberidae

Identification: 2 to 3 inches long, wingless, brown abdomen, black head and thorax

Distribution: Native to Madagascar; sold on eBay, Amazon, and other websites

Comments: The hissing sound that male Madagascar hissing cockroaches make when they are disturbed is produced by releasing air through modified spiracles. It is used to attract mates and chase other males from their territory. These cockroaches are kept as pets and as food for pet reptiles throughout North America. You might wonder why anyone would keep Madagascar hissing cockroaches as pets. Well, watching them is interesting, and they are easy to maintain, cannot fly or bite, and if they escape they will not survive to infest your home or apartment. In their native Madagascar, they are found in logs. Madagascar hissing cockroaches are ovoviviparous, which means females give birth to live nymphs.

Female Madagascar hissing cockroach giving birth to live nymphs

16 MANTISES

Mantises are ambush predators that feed on both beneficial insects as well as pests. Thus, whether they can be considered beneficial is debatable. Their front legs are armed with spines for grasping prey. It has been reported that female mantises sometimes eat their partner after they have mated. However, it has also been argued that they may only eat their partner when they are in captivity. Even though devouring your mate may seem strange to us, it may be of selective advantage for a male mantis to be consumed by his mate. It is known that when females of many kinds of animals, including insects, are well-fed, they produce more and healthier young than if they are undernourished. Thus, being consumed by his mate may be of selective advantage to a male, especially if he is near the end of his life, as this will ensure that his mate is well-fed and thus will produce more and healthier offspring. They are his young too.

CHINESE MANTIS, CHINESE PRAYING MANTIS

Latin name: *Tenodera sinensis*

Family: Mantidae

Identification: 4.5 inches long, brown or green, distinguished from other mantises by a spot on their front legs

Distribution: Throughout North America, especially in the Northeast

Comments: Chinese mantises are the largest mantises in North America. They were first introduced to Mount Avery, Pennsylvania, in 1896 by a gardener and have been imported many times since. The egg cases of Chinese mantises are sold to gardeners to control insect pests and hobbyists who keep them as pets. Like other species of mantises, Chinese mantises hold their front raptor-like legs in a position reminiscent of a praying person. Being an ambush predator, the Chinese mantis is praying for an insect to walk by so that he or she can grab and devour it. They attack any prey that is small enough to take on, from small grasshopper nymphs to hummingbirds. Although many gardeners purchase them or their eggs to put in the garden as a biological control, their status as beneficial insects is debated because a Chinese mantis would just as soon eat a beneficial insect as a pest.

EUROPEAN MANTIS, PRAYING MANTIS

Latin name: *Mantis religiosa*

Family: Mantidae

Identification: 3 inches long, black marking inside front legs

Distribution: Throughout eastern North America

Comments: European mantises were accidentally introduced in 1899 on plants and have spread throughout the Northeast. Like other mantises, the European mantis is an ambush hunter. They locate their prey with their large eyes, which each has over 10,000 ommatidia and a small area called the fovea, which has greater resolution than the rest of the eye. They are able to turn their head to look directly at an object without moving the rest of their body. In fact, they are the only insects that can look over their shoulders.

Mantises have many predators, especially birds, small mammals, and bats. Their greatest defense is their camouflaged coloration that blends in with the plants where they live and hunt. An earlike organ on their thorax allows them to hear the echolocation sounds of bats. If they hear an approaching bat when they are flying, they immediately land, and if they are perched, they remain motionless.

17 STICK INSECTS

Stick insects are well-named: They resemble sticks. They are nocturnal insects that feed on the leaves of trees on which they live. Walking sticks are one of the few insects that can change color. They change from green to brown and back again under the influence of light, temperature, or humidity. Usually, they are dark brown at night and green during the day to match the color of the leaves that they live and feed on. Despite their excellent camouflage, walking sticks are not common, probably because they reproduce slowly.

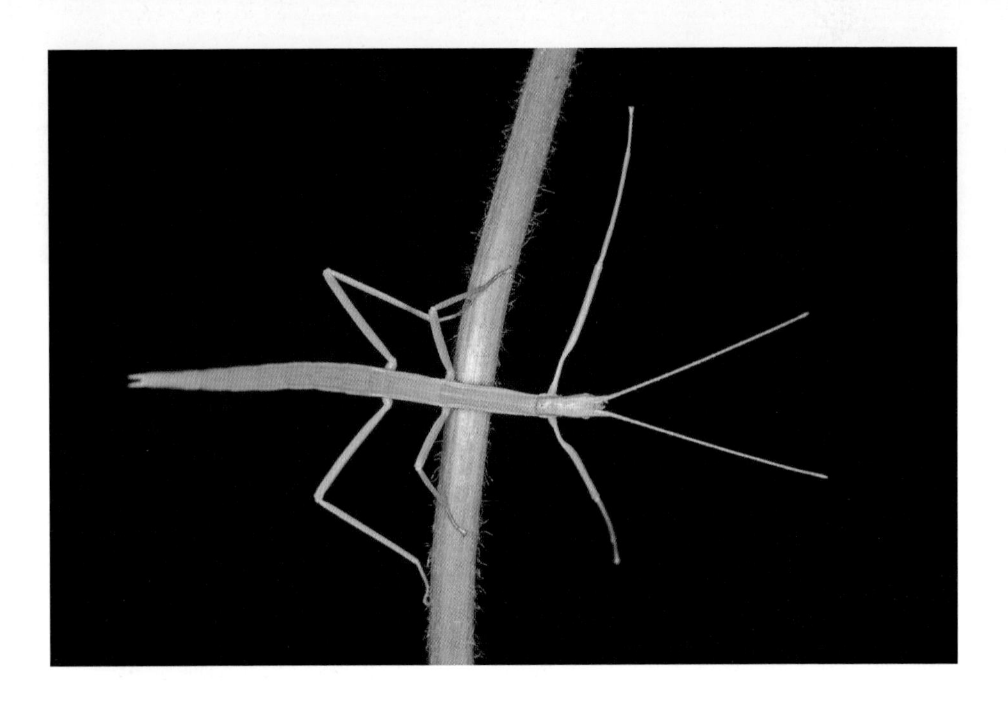

NORTHERN WALKING STICK, STICK INSECT

Latin name: *Diapheromera femorata*

Family: Phasmidae

Identification: 3 to 3.5 inches long, slender, green or brown, looks like a stick with 6 legs

Distribution: East Coast from Maine to Florida, west to New Mexico, north to North Dakota; Canada from Quebec to Alberta

Comments: Although some tropical species, and one that lives in southern Florida, can fly, the other species of stick insects that live in North America are wingless. Stick insects are not easy to find because they usually live high up in trees. During the day, these slow-moving nocturnal insects avoid detection by sitting motionless or waving a bit like a stick might move in the wind, then at night, they feed on leaves. When attacked, walking sticks may fall to the ground and remain motionless for some time. They can also secrete a foul-smelling substance from their thorax to dissuade predators.

Unlike most insects, young stick insects are usually able to regenerate lost limbs. *Diapheromera femorata* and most other species of stick insects reproduce by parthenogenesis, although a few males can sometimes be found. Eggs fall to the ground and hatch the second year after they are laid.

18 GRASSHOPPERS

Grasshoppers, as well as crickets, are known for their powerful hind legs that they use to hop from danger and their sound-producing organs. To avoid detection, most species are camouflaged to blend in with the grasses on which they live and feed. Many species can also secrete a noxious fluid from their mouth to deter predators. Grasshoppers use their chewing mouthparts to feed on various grasses, including wheat, alfalfa, corn, rice, and other grass crops, and they are serious pests of these crops. They are either short-horned grasshoppers (family Acrididae) that have relatively short antennae or long-horned grasshoppers (family Tettigoniidae) that have long, thin, tapering antennae. Long-horned grasshoppers are also called katydids or bush crickets.

Crickets can be distinguished from grasshoppers by how they sing (stridulate). Grasshoppers stridulate by rubbing their hind legs against their wings, while crickets rub their wings together. The hearing organs of grasshoppers are at the base of their abdomen, while those of crickets are on their front legs. Also, most grasshoppers are diurnal, whereas most crickets are crepuscular. Most grasshoppers feed on grasses, but crickets tend to be omnivorous. They often feed on decaying animal and plant matter and small, soft-bodied insects.

TWO-STRIPED GRASSHOPPER

Latin name: *Melanoplus bivittatus*

Family: Acrididae

Identification: 0.13 to over 2 inches long, pale yellow stripes running down back

Distribution: Throughout the United States and southern Canada

Comments: Grasshoppers of the genus *Melanoplus* are called spur-throated grasshoppers, named for the peg-like process that protrudes from their "neck." They feed on all sorts of grasses and are some of the most serious pests in North America. Two-striped grasshoppers are particularly destructive pests in the Midwest and southern Canada, where wheat, alfalfa, oats, corn, rye, rice, and other grasses are farmed.

Locusts are actually grasshoppers, many of which belong to the genus *Melanoplus*. When these solitary grasshoppers are overcrowded or there is a drought, the young nymphs become more sociable, feed on poisonous plants, and become more colorful. After they molt into adults, they fly off in giant migratory swarms. They are then no longer called grasshoppers but locusts. From 1874 to 1877, swarms of Rocky Mountain locust (*Melanoplus spretus*) ravaged farms in the Great Plains. It would be nice to show the reader a photograph, but this species of *Melanoplus* has been extinct for over a hundred years.

PINE TREE SPUR-THROATED GRASSHOPPER, GRIZZLY SPUR-THROATED GRASSHOPPER

Latin name: *Melanoplus punctulatus*

Family: Acrididae

Identification: 0.7 to 1 inch long, gray and black

Distribution: Eastern and central United States, southeastern Canada (uncommon)

Comments: Like other spur-throated grasshoppers, pine tree spur-throated grasshoppers feed on grasses. Although spur-throated grasshoppers are some of the worst pests in North America, pine tree spur-throated grasshoppers are less so because they are relatively uncommon and, therefore, not much of a problem where grass crops are farmed. They are well camouflaged to blend in with pine trees and the leaf litter under the trees where they live.

NORTHERN GREEN-STRIPED GRASSHOPPER

Latin name: *Chortophaga viridifasciata*

Family: Acrididae

Identification: 1 to 1.5 inches long, green, short antennae, brown forewings and legs

Distribution: Throughout the Northeast to the Rocky Mountains, southern Canada

Comments: Most grasshopper overwinter as eggs, but green-striped grasshoppers overwinter as nymphs. Because of this, they are usually the first grasshopper to be seen in the spring. However, being short-lived, by the middle of the summer green-striped grasshoppers are mostly absent from the meadows and roadside grasses that they inhabit. If you're looking for one of these grasshoppers, you might want to look over your shoulder, because if they are on a blade of grass, they often move around to the opposite side when they detect someone approaching. Even in eastern North America, where they are most common, green-striped grasshoppers are never in high enough numbers to be considered major agricultural pests.

PIGMY GRASSHOPPERS, GROUSE GRASSHOPPERS

Latin name: *Paratettix* sp.

Family: Tetrigidae

Identification: 0.5 to 0.7 inch long, usually brown, pronotum extends over abdomen and usually ends in a point

Distribution: Different species throughout North America

Comments: Unlike most species of short-horned grasshoppers that overwinter as eggs, pigmy grasshoppers overwinter as adults and can, therefore, be found in the spring and early summer. Most species live along the shores of ponds and streams, where both nymphs and adults feed on algae. Although pigmy grasshoppers can easily be distinguished from other short-horned grasshoppers by their relatively small size and long pronotum, it is often difficult to identify their species because members of the same species can have either short or long wings or can be wingless. Also, the color patterns can vary among individuals of the same species, and males and females can have different color patterns. Females are larger than males. Although most species are brown, they sometimes appear greenish because they have algae growing on them.

EASTERN SHIELDBACKS

Latin name: *Atlanticus* sp.

Family: Tettigoniidae

Identification: 0.8 to 1.4 inches long, brown with short pronotum curving upward posteriorly

Distribution: Eastern United States

Comments: Shieldbacks are long-horned grasshoppers. They can be distinguished from other long-horned grasshoppers by their pronotum, which curves upward such that it is somewhat reminiscent of a shield. Although shieldback katydids are found across the United States, they are most common in California. One California species, the Antioch Dune shieldback katydid (*Neduba extincta*), was the largest katydid in North America. Unfortunately, the species was driven to extinction due to loss of habitat. Like most other Tettigoniids, male shieldbacks have sound-producing organs on their front legs that can produce a continuous song, known as a trill, to call mates.

TRUE KATYDID, NORTHERN TRUE KATYDID, NORTHERN KATYDID

Latin name: *Pterophylla camellifolia*

Family: Tettigoniidae

Identification: 1.7 to 2.4 inches long, green, oval forewings that look like leaves

Distribution: Massachusetts south to northern Florida, west to Iowa, south to eastern Texas

Comments: The true katydid is the species that calls *katy-did, she-did katy-didn't*, from which the Tettigoniidae get the name katydid. Males call to attract mates, and females reply with a scraping sound. Calls of many insects are made by rubbing two parts together, such that one part vibrates, emitting a sound similar to the musical sound of a violin's vibrating strings. In the case of the true katydid, the sound is produced by rubbing their forewings together. One wing has a sawtooth edge called a file, and the other wing has a structure called a scraper. The sound is amplified by their wings. True katydids are not easy to see because their green wings resemble the leaves of the trees and bushes on which they live and feed. Also, they cannot fly.

FORKED-TAILED BUSH KATYDID, FORKED-TAILED KATYDID

Latin name: *Scudderia furcata*

Family: Tettigoniidae

Identification: 1.6 to 2 inches long, narrow green forewings, females have pink ovipositor, males have forked claspers at tip of abdomen

Distribution: Throughout the United States and southern Canada

Comments: Male forked-trailed bush katydids employ their fork-shaped claspers at the tip of their abdomen to hold on to females during mating. These katydids feed on the leaves of many kinds of plants. In California, *Scudderia furcata* sometimes damages young oranges by feeding on the rind. Male forked-tailed katydids call females by making two to four short chirps, and females chirp in response. In late summer or early fall, females attach their eggs to leaves and twigs. Eggs overwinter and hatch in the spring.

LESSER ANGLE-WING, BROAD-WINGED KATYDID

Latin name: *Microcentrum retinerve*

Family: Tettigoniidae

Identification: 2 to 2.5 inches long, green

Distribution: Throughout the United States and southern Canada

Comments: Lesser angle-wing katydids are similar to their cousins, the greater angle-wings (*Microcentrum rhombifolium*), which are a bit larger. The lesser angle-wings are usually seen from July until October. They feed on the leaves of a number of different kinds of trees. Lesser angle-wings are somewhat similar to the true katydid, *Pterophylla camellifolia,* but can be distinguished by their pointed hindwing protruding beyond the forewings. During mating, the male transfers sperm as well as food to the female to ensure that their young will develop into healthy young nymphs. The protein-rich food can be a third of the body weight of the male. The lesser angle-wing's call, which is made by both sexes, sounds a bit like two pebbles tapping together. Females lay their eggs in single rows on the margins of leaves or in double rows on stems.

Male meadow grasshopper

MEADOW KATYDIDS, LONG-HORNED GRASSHOPPERS

Latin name: *Conocephalus* sp.

Family: Tettigoniidae

Identification: 0.8 to 1.2 inches long, green or brown, long thin antennae, large upward-curved ovipositor, males with prominent cerci

Distribution: Eastern United States and southeastern Canada

Comments: Meadow katydids are common in fields and grasslands. Different species of the genus *Conocephalus* are often identified by the shape of the two curved cerci on the abdomen of males. Meadow katydids do not move around much during the day, but when night comes, they forage for food. Females lay eggs in the fall, and they are fussy about where they place them. Typically females chew holes in many plants until they find one that suits them. They then turn around and use their long ovipositor to deposit eggs in the hole, before chewing the hole back together. The eggs overwinter, and the young nymphs emerge in the spring. To call females, male meadow katydids produce high-pitched sounds composed of rattles, buzzes, and ticks, which sound very different from the chirps of crickets. To detect the calls, both male and female meadow katydids have a hearing organ, called a tympanum, on their front legs.

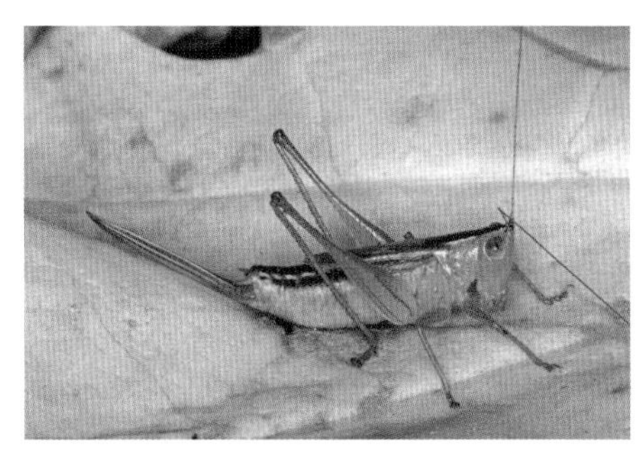

Female meadow grasshopper

19
CRICKETS

Crickets are small to medium-size insects with cylindrical bodies and round heads. Males have two long cerci, and females have prominent ovipositors. Male crickets can make different songs to either call mates, encourage their mate to lay eggs, or frighten off other males. When they hatch from their eggs, nymphs undergo about ten molts before they become adults. Newly hatched adults take a few days before they can mate.

SPOTTED CAMEL CRICKET, CAVE CRICKET, SPIDER CRICKET

Latin name: *Ceuthophilus maculatus*

Family: Rhaphidophoridae

Identification: 1.5 to 2 inches long, brownish, humpbacked, long thin antennae

Distribution: Throughout eastern United States and southern Quebec and Ontario

Comments: Camel crickets get their name because they are humpbacked. They are also called cave crickets. Although they do live in caves, these flightless nocturnal crickets can also be found in rotten logs, hollow trees, under rocks and floorboards, in cellars, and in other damp places. In cellars, they can be a nuisance, although they are harmless. Camel crickets feed on decaying vegetable matter and plants. An Asian camel cricket species, *Diestrammena asynamora,* which was introduced in the nineteenth century, has spread throughout much of North America and is more common in some areas than the native species of camel crickets.

FIELD CRICKETS, FALL FIELD CRICKETS

Latin name: *Gryllus* sp.

Family: Gryllidae

Identification: 0.6 to 1 inch long, brown or black, 2 prominent cerci and ovipositor

Distribution: Southern Canada, most of the United States, absent from the Southeast

Comments: Male field crickets employ their song to attract mates. Females can detect whether a male is healthy by its sound and on this basis decide if she wants to mate with him. When the female approaches a male, he produces a softer courtship chirp. After mating, males often remain with females to ensure that they do not mate with other males. Although the chirping of males carries for some distance and attracts mates, insects that broadcast their position may attract less-welcome visitors. Females of a small fly, *Ormia ochracea*, lay their eggs on field crickets. When the fly larvae hatch, they feed on the crickets, which ultimately kills them. These tiny flies have an auditory organ that has been shown to enable female flies to follow the sound of a cricket with an error of only 2 degrees.

HOUSE CRICKET

Latin name: *Acheta domesticus*

Family: Gryllidae

Identification: 0.5 to 0.8 inch long, straw color with reddish-brown markings

Distribution: British Columbia east to Newfoundland and throughout the United States except Florida

Comments: House crickets are native to Asia and North Africa. They are the crickets of the Dickens novel *The Cricket on the Hearth* and George Selden's *The Cricket in Times Square.* In parts of Asia, house crickets are raised for food, and in North America, they are reared as food for pet amphibians and reptiles. In the winter, males of these nocturnal crickets sometimes enter homes, where they chirp at night. Depending on your point of view, their song is either beautiful music or an annoyance. In ancient China, men carried a little cage with a male house cricket inside. The men would put the crickets together and bet money on whose cricket would win the fight.

SPHAGNUM CRICKET, SPHAGNUM GROUND CRICKET, MARSH GROUND CRICKET

Latin name: *Neonemobius palustris*

Family: Gryllidae

Identification: 0.2 to 0.4 inch long, usually black but can be brown or tan

Distribution: Eastern United States, west to Minnesota and Arkansas, absent from southern Florida; southeastern Canada

Comments: Sphagnum crickets feed exclusively on sphagnum, which are mosses that grow on the edges of ponds, lakes, rivers, and other wet places. These mosses, which are sometimes called peat moss, can store water up to 25 times their dry weight. As female crickets also lay their eggs on sphagnum, the best place to find these little crickets is on sphagnum moss. Sphagnum crickets hide in the wet moss and can be found by pushing the moss down, which often causes the crickets to hop out. They can also be located by their song, which is composed of high trills that start softly and become louder. Each trill lasts about 10 seconds and is separated from the next trill by several seconds of silence.

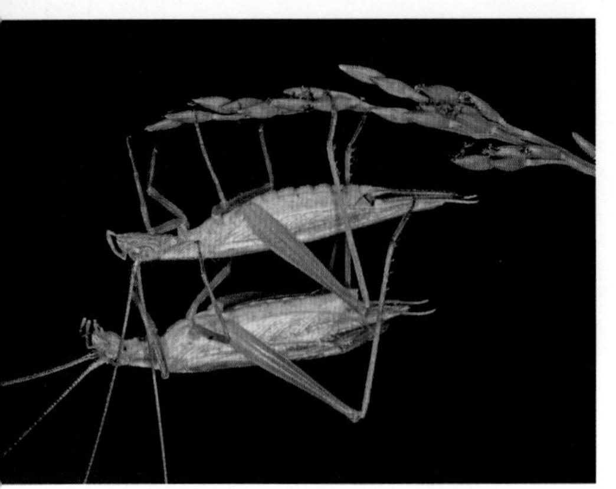

A pair of snowy tree crickets, female above

SNOWY TREE CRICKET, THERMOMETER CRICKET

Latin name: *Oecanthus fultoni*

Family: Gryllidae

Identification: 0.4 to 0.5 inch long, pale green, transparent wings

Distribution: Much of the United States and southern Canada, absent from southern states

Comments: "Tree cricket" is probably not the best name for these crickets because they also live in bushes and fields. This is good from my point of view because it is difficult to climb a tree carrying a net or a camera. Snowy tree crickets are common across most of North America. Although many people have never seen one, their beautiful chirps are familiar to most people in North America who live near a field or wooded area. Nathaniel Hawthorne explained it as "audible stillness" and wrote, "If moonlight could be heard, it would sound just like that." The chirping is made by rubbing the ridges of their wings together. Only males can chirp. Snowy tree crickets are known as "thermometer crickets" because the number of chirps in a given amount of time depends on the temperature. The temperature in Fahrenheit can be estimated by counting the chirps in 13 seconds and adding 40. The formula varies a bit from location to location.

TWO-SPOTTED TREE CRICKET

Latin name: *Neoxabea bipunctata*

Family: Gryllidae

Identification: 0.4 to 0.5 inch long, females have a brown head and thorax with brown and white forewings, males have a light reddish-brown head and thorax and clear forewings

Distribution: Massachusetts south to northern Florida, west to Oklahoma

Comments: Two-spotted tree crickets are common in North America, especially in the East. They are brown or gray and have two dark spots on their back. All tree crickets display what is termed "courtship feeding." After mating, the cricket produces a secretion from a gland between its thorax and wings and allows his mate to feed on the fluid. The fluid contains nutrients that help the female produce more and healthier eggs. This is of selective advantage to the male as well as the female, as they are his offspring too. Females lay their eggs in the fall in holes that they drill in bark with their ovipositor. The eggs hatch in the spring, and the young nymph tree crickets begin feeding, usually on tiny insects, especially aphids.

20 HEMIPTERA

Hemipterans are characterized by piercing and sucking mouthparts that are contained in a sheath called a rostrum or proboscis. The proboscis opens lengthwise along the front, revealing the two long, thin, needle-like mouthparts called stylets, which are designed for piercing. The rear two mouthparts are fused into a structure containing two channels: a small channel through which the bug can secrete either salivary gland enzymes or toxins, and a larger channel through which the bug pumps up liquid food.

Most hemipterans feed on the sap of plants. When they feed, they first pierce the plant. They then inject enzymes that partially digest the plant tissue before they pump up the sap. Hemipterans that feed on animals use their front stylets to wound the animal before they inject toxins that paralyze it or enzymes that partially digest the tissue at the wound site before they suck up those tissues.

The forewings of hemipterans are referred to as hemelytra because they are thickened at the base and membranous in the apex. They protect the abdomen and beat when the insect flies.

Aphids

Aphids are tiny, soft-bodied hemipterans that reproduce rapidly and are an important source of food for many kinds of insects. They are also serious pests of many garden plants, houseplants, and agricultural crops. Unlike most hemipterans, aphids have the ability to tap directly into the xylem channels that carry water and nutrients, or sap, upward from the roots of the plant to the leaves and the phloem tissue that transports sap from leaves to other tissues.

GOLDENGLOW APHIDS

Latin name: *Uroleucon* sp.

Family: Aphididae

Identification: 0.08 to 0.1 inch long, red with cornicles

Distribution: Throughout North America

Comments: Goldenglow aphids can usually be found in fields and gardens, where they feed on a number of plants. These and other aphids have two tubes protruding from their abdomen called cornicles. When attacked, aphids secrete defensive wax through their cornicles. The wax sticks to predators and quickly hardens. Consequently, small predatory wasps that attack aphids can often be found dead, stuck to plants where aphids are living. Some individuals are wingless, but others in the same group have wings. Although these tiny soft-bodied insects can't walk or fly very far, they can travel long distances on the wind. Aphids, like many other insects, are also spread by man. In fact, it is difficult to stop the spread of aphids since people transport plants for food and gardens, as well as numerous other items, with these tiny insects often going along for the ride undetected.

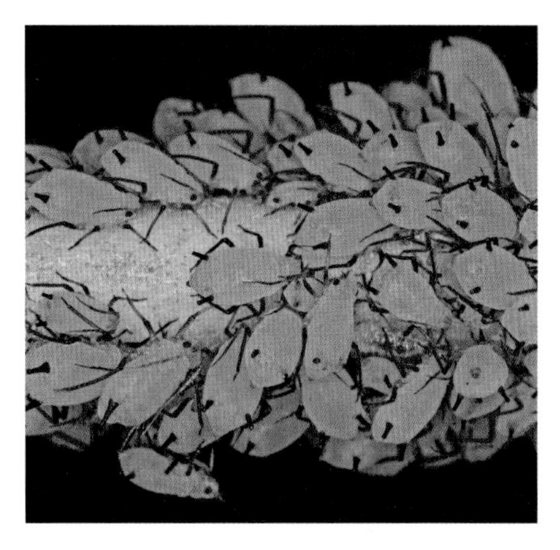

OLEANDER APHID, MILKWEED APHID

Latin name: *Aphis nerii*

Family: Aphididae

Identification: 0.08 to 0.1 inch long, yellow or orange with black cornicles, legs, and antennae

Distribution: Throughout North America

Comments: In addition to milkweed, this aphid species feeds on a number of important ornamental flowering plants. Although this aphid probably originated in the North African or Mediterranean region where oleander is from, it is now found all over the world, including all of North America. Oleander aphids are all parthenogenetic females and can have wings or be wingless. Nymphs are similar to adults. Plants that are infested by these aphids are often stunted and unsightly because the aphids feed on the sap from the plants. These aphids sequester poisons from the poisonous oleander and milkweed plants that they feed on and are thus avoided by many predators. Their yellow or orange warning color forecasts to predators that they are poisonous.

CRAPEMYRTLE APHID

Latin name: *Tinocallis kahawaluokalani*

Family: Aphididae

Identification: 0.08 to 0.1 inch long, pale yellow

Distribution: Throughout the southern United States and wherever crape myrtle grows

Comments: Crapemyrtle aphids are said to be ovoviviparous because females produce eggs that hatch within their body and thus give birth to live young. Sharks, rays, and some other species of fish, amphibians, and reptiles, as well as some invertebrates, are also ovoviviparous. Crapemyrtle aphids, like other aphid species, typically produce many generations in a single season. In some aphid species, all the individuals are parthenogenetic ovoviviparous females. However, in this and many other species, individuals can be either parthenogenetic females or sexual females or males, depending on the season. In these species, the young that are produced in the spring become parthenogenetic females. These females produce parthenogenetic females that continue for a number of generations until the fall when the parthenogenetic females give birth to fertile males and females. These sexual aphids mate, but the females are not ovoviviparous or parthenogenetic. Instead, they lay eggs that overwinter and become parthenogenetic females when they hatch in the spring. Thus, these aphids enjoy the benefits of both parthenogenetic and sexual reproduction.

DOG DAY CICADA

Latin name: *Tibicen canicularis*

Family: Cicadidae

Identification: 1 to 1.3 inches long, black with green markings, gray eyes

Distribution: Northern United States and southern Canada, east of the Rocky Mountains

Comments: Dog day cicadas can best be distinguished from periodical cicadas (17-year or 13-year locust) by their gray eyes. Periodical cicadas have red eyes. Cicadas are probably best known for the buzzing calls that males produce to attract females, which are often described as sounding like the noise of a power saw cutting wood. The sound is produced by a pair of organs underneath the male's abdomen. If you would like to hear it, pick up a male cicada, and it will usually sound the call.

Adult cicadas do not feed. Females lay their eggs in slits that they make in twigs with their ovipositor. The injury can cause the twig to break off, but the damage is usually minor. The larvae fall to the ground, where they burrow into the soil and feed on sap from tree roots for 3 to 5 years before they emerge, climb trees, grasp the tree tightly, and undergo their adult molt.

The dog day cicada on the right has just molted. Its wings have not yet hardened. The empty brown exoskeleton is above.

THORN-MIMIC TREEHOPPER

Latin name: *Enchenopa* sp.

Family: Membracidae

Identification: 0.2- to 0.5-inch-long body, enlarged pointed pronotum

Distribution: Throughout North America

Comments: Like most treehoppers, those of the genus *Enchenopa* resemble thorns. They usually perch on twigs and branches of the plants on which they feed, disguised as thorns. Individuals of some species line up on small branches, all facing the same way, so as to appear as a group of thorns. These hemipterans communicate with one another by tapping on the tree branch. Like most other hemipterans, both adults and nymphs feed on sap and secrete the water and sugar, known as honeydew, that they do not consume from their anus. Nymphs have a long tube that they can extend from their anus to shoot honeydew away from their bodies. It is important for plant-feeding hemipterans to dispose of honeydew because it can be infected with molds that can kill them. Most species of treehoppers are innocuous to humans, but a few are minor pests.

When viewed head-on, the buffalo treehopper is reminiscent of an American bison.

BUFFALO TREEHOPPERS

Latin name: *Stictocephala* sp.

Family: Membracidae

Identification: 0.2 to 0.3 inch long, stubby, bright green

Distribution: Throughout North America

Comments: Viewed from the front, these stubby little green treehoppers are reminiscent of American bison. Males attract females with calls that are too high-frequency for us to hear. Most treehoppers do not cause significant damage, with the exception of the buffalo treehopper, which often causes serious damage to fruit trees, especially apple trees. Females lay their eggs in slits in the twigs of apple trees that they make with their knifelike ovipositor. Frequently, the terminal portion of the twig beyond where the eggs are placed dies. The eggs overwinter and hatch in the spring, and the nymphs fall to the ground, where they feed on various weeds and non-woody plants before returning to the trees.

The forewings of buffalo treehoppers can be seen when they are viewed from the side.

CANDY-STRIPED LEAFHOPPER

Latin name: *Graphocephala coccinea*

Family: Cicadellidae

Identification: 0.1 to 0.6 inch long, wedge-shaped, large back legs with a row of spines on lower part of legs

Distribution: Throughout North America

Comments: About 2,500 species of leafhoppers have been described in North America. The most obvious leafhoppers to most people are the ones that hop around on lawns, where they feed on the sap from grasses. *Graphocephala coccinea* is one of many species that are quite colorful, but because they are small insects, it is difficult to see their color patterns without a magnifying glass. The prominent rows of little spines on their back legs, which they employ for traction when they hop, distinguish leafhoppers from closely related hemipterans. Although most species do not cause significant damage, some are major pests of cultivated plants. In addition to damage caused by the feeding, leafhoppers can transmit bacterial and viral diseases that damage or kill the plants on which they feed and can spread from one plant to another. One of the most serious agricultural pests is the beet leafhopper (*Circulifer tenellus*), which can transmit the beet curly top virus to tobacco, tomatoes, and eggplants.

TWO-LINED FROGHOPPER, TWO-LINED SPITTLEBUG

Latin name: *Prosapia bicincta*

Family: Cercopidae

Identification: 0.3 to 0.4 inch long, black with 2 yellow, orange, or red lines crossing wings and a dull red line on pronotum

Distribution: Massachusetts to Florida, west to Kansas and Texas

Comments: Two-lined froghoppers live in fields and gardens, where they feed on the sap of grasses, grains, and other plants. The bright orange, yellow, or red lines on the forewings of these froghoppers are warning colors. They remind predators that the froghopper produces unpleasant-smelling defense chemicals. When they feed, two-lined froghoppers inject a toxin that causes the grasses to produce less chlorophyll and turn brown. Like other species of froghoppers, nymphs produce a spittle mass sometimes called "cuckoo spit." Large amounts of cuckoo spit can cause farm machinery to clog due to spit-wettened plants and stems becoming entangled in the machinery.

RED-LEGGED FROGHOPPER, RED-LEGGED SPITTLEBUG

Latin name: *Prosapia ignipectus*

Family: Cercopidae

Identification: 0.3 to 0.4 inch long, black with red and black legs

Distribution: Northeastern United States, west to Missouri and Iowa

Comments: Red-legged froghoppers usually feed on grasses, including centipede grass, Bermuda grass, and corn. These little hoppers are called froghoppers because some people think they resemble minute frogs. Actually, they resemble leafhoppers. The two can be distinguished from one another because froghoppers have only a few stout spines on their hind legs, whereas leafhoppers have many small sharp spines. In both cases, the spines are used to dig in for traction as they leap into the air. Although they both can fly, leaping is the most common means of escape for froghoppers and leafhoppers. In fact, for their size, a good hop is the equivalent of a person jumping the length of a football field.

BROAD-HEADED SHARPSHOOTER

Latin name: *Oncometopia orbona*

Family: Cicadellidae

Identification: 0.5 to 0.6 inch long, reddish brown with numerous small cream-colored spots, large protruding eyes

Distribution: Eastern United States and southeastern Canada

Comments: Sharpshooters are large leafhoppers, and like other insects in the Cicadellidae family, they have piercing and sucking mouthparts. They extract nutrients and squirt the unwanted honeydew from their anus, which is how they got the name sharpshooter. Although they are excellent jumpers and good fliers, when threatened, they often quickly move behind branches or stems to avoid detection. Females use their sharp, knifelike ovipositor to inject eggs into plant stems. They produce a chalklike substance over their wings, which they scrape off and use to cover their eggs for protection. When the eggs hatch, the nymphs undergo five molts. Broad-headed sharpshooters have one generation a year.

This meadow froghopper nymph was hiding in bubbles. I teased him out.

MEADOW FROGHOPPER, MEADOW SPITTLEBUG

Latin name: *Philaenus spumarius*

Family: Aphrophoridae

Identification: 0.2 to 0.3 inch long, multiple colors and patterns

Distribution: Northern North America and southern Canada

Comments: You can't identify *Philaenus spumarius* by their color or color pattern because they come in at least twenty variations. Froghoppers are also called spittlebugs because their nymphs produce a frothy substance in which they live. Like many of their relatives, such as aphids and treehoppers, froghoppers suck up sap from plants, retain nutrients, and expel the unwanted honeydew from their anus. The substance that froghopper nymphs produce contains an enzyme that digests a wax that is produced by their anal gland. This creates a unique soap-like fluid that froghopper nymphs let accumulate over their bodies and the plant stem on which they are situated. When enough of the soap is produced, the nymph dips the tip of its abdomen into the soap and lets air escape to form a bubble. This is repeated every second or so until the nymph is covered with bubbles. It is believed that the frothy mass provides them with a safe hideaway from predators and keeps them moist.

Meadow froghopper. One of twenty different color patterns

CITRUS FLATID PLANTHOPPER

Latin name: *Metcalfa pruinosa*

Family: Flatidae

Identification: 0.2 to 0.3 inch long, whitish, sometimes covered with wax, yellow or orange eyes

Distribution: Ontario and Quebec, south to Florida, west to California

Comments: Some imported pests cause a great deal of damage in North America. However, just as imported insects can harm North American plants, harmful native North American pests can travel to other countries. The citrus flatid planthopper is a good example. Although they are widely distributed in North America and feed on numerous plants including citrus and other crops, these planthoppers rarely cause significant damage, presumably because their numbers are controlled by parasitoid wasps and other predators. In 1979, citrus flatid planthoppers were accidentally introduced into Italy. Since then they have spread throughout south and central Europe and Russia, where they cause significant damage to many flowering shrubs and commercial crops, especially grapes and various other fruits.

Most two-striped planthoppers are green.

TWO-STRIPED PLANTHOPPER

Latin name: *Acanalonia bivittata*

Family: Acanaloniidae

Identification: 0.15 to 0.3 inch long, resembles a green or red leaf, 2 cylindrical brown stripes on top of forewings, brown head and thorax

Distribution: Throughout the United States and southern Ontario, absent from the West Coast

Comments: Two-striped planthoppers are convincingly camouflaged as leaves. Most look like green leaves, but some appear like orange/red leaves. Don't be fooled by color. Slight changes in the structure of colored molecules can cause them to reflect entirely different colors. These planthoppers feed on a variety of plants but are usually not very numerous and, therefore, are not considered major pests. Females produce a wax to cover and protect their eggs. Nymphs also produce a wax that coats and helps prevent desiccation. The wax also enables nymphs to float if they fall into a puddle and helps prevent them from sticking to spiderwebs.

Some two-striped planthoppers are orange.

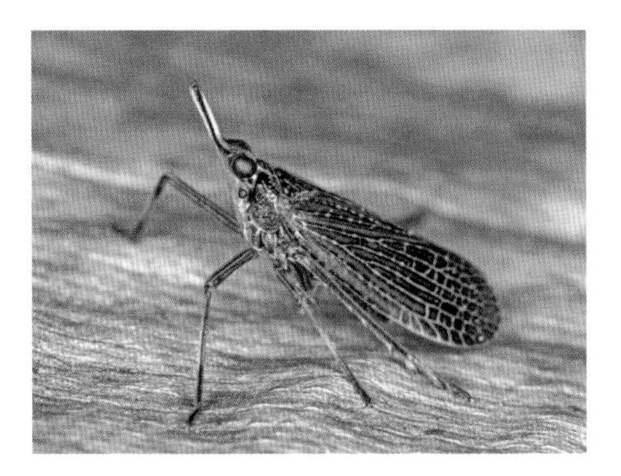

PARTRIDGE SCOLOPS, PARTRIDGE BUGS

Latin name: *Scolops* sp.

Family: Miridae

Identification: 0.2 to 0.3 inch long, long nose

Distribution: Northeastern United States and southern Canada, west to Missouri

Comments: These little plant bugs are common in fields, prairies, and the edges of forests across much of North America. They have a distinctive shape, with a long nose called a frons that points up when the bugs are resting. If you have seen a picture of one of these distinctive little bugs and then see one in a meadow, you will know immediately that it is a partridge scolops. Like most plant bugs, they are soft-bodied insects that use their mouthparts to inject enzymes into plants and pump up the partly digested tissue.

MEADOW PLANT BUG

Latin name: *Leptopterna dolabrata*

Family: Miridae

Identification: 0.3 to 0.5 inch long, orange and yellow elytra, black head and thorax with yellow or orange markings, black legs and antennae, males have long wings, females have short wings or wingless

Distribution: Maritime Provinces to North Carolina, west to Nebraska, southern Canada, occasionally West Coast

Comments: Meadow plant bugs were accidentally imported from Spain, where they are common. They feed on grass seeds, causing the seeds to shrivel and turn white. They are considered pests because they feed on grass crops, as well as wild grasses. Meadow plant bugs are especially fond of Kentucky bluegrass, which is common throughout North America. When they feed, these plant bugs release a sticky fluid that can break down gluten in wheat, which in turn has a deleterious effect on flour. Females use their ovipositor to bore holes near the base of grass stems, causing injury to the stems. The eggs that plant bugs deposit in the stems overwinter and hatch in the spring.

PLANT BUGS, LEAF BUGS, CAPSID BUGS, GRASS BUGS

Latin name: *Lopidae* sp.

Family: Miridae

Identification: 0.2 to 0.3 inch long, orange pronotum, orange and black wings, humpbacked, dark head

Distribution: Across North America

Comments: Plant bugs are a large and diverse group of hemipterans. Almost all are plant feeders, and some are pests in gardens and on crops. However, most, like *Lopidae* sp., live in fields and woodlands and do not cause harm to gardens or crops. Many species are referred to as Holarctic, which means that the same species live across the northern part of the globe, including North America, North Africa, and northern Eurasia. The reason that some insects, as well as other animals, live throughout these regions may have to do with the fact that the glaciers that covered all of the regions created similar ecosystems. Also, all of these regions were connected by the Bering land bridge that existed between Siberia and Alaska up to some 15,000 years ago.

SHORE BUGS

Latin name: *Salda* sp.

Family: Saldidae

Identification: 0.2 to 0.3 inch long, oval-shaped, brown, prominent front legs, large protruding eyes, long closed cells on hemelytra

Distribution: Throughout the United States and southern Canada

Comments: These unusual-looking little insects live along banks of ponds, lakes, rivers, salt marshes, and mudflats. They tend to move rapidly, running, leaping, and taking short flights as they look for little insects on which they feed. When disturbed they usually fly or leap and then hide under vegetation or in crevices. Some species burrow in mud along shores of their habitat.

SMALL MILKWEED BUG

Latin name: *Lygaeus kalmii*

Family: Lygaeidae

Identification: 0.4 inch long, red and black markings, red spot on head

Distribution: Most of the United States and southern Canada, rare in Florida

Comments: The small milkweed bug feeds on seedpods of the milkweed plant, so that's where you usually find them. Occasionally they will fly off some distance from the milkweed. They sometimes suck liquid from dead insects and feed on the eggs of swamp milkweed leaf beetles. They are immune to the poisonous milkweed plant, but like other insects that feed on milkweed, they sequester the plant's poison. Their bright red color warns predators that they are poisonous.

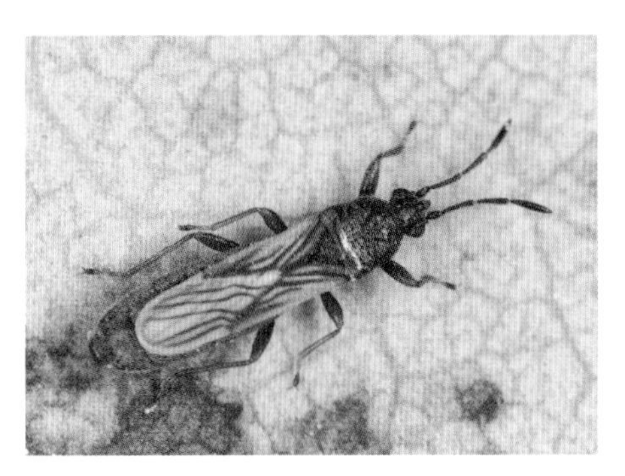

SALTMARSH CHINCH BUG

Latin name: *Ischnodemus falicus*

Family: Lygaeidae

Identification: 0.2 to 0.3 inch long, thin and flat, brown, distinctive cream-colored elytra with thin brown lines

Distribution: Eastern United States and southeastern Canada

Comments: Saltmarsh chinch bugs are found in grasses along salt marshes, ponds, and lakes. This species does not cause much damage to the grasses on which they feed, but other chinch bug species can cause significant damage to grass crops including barley, wheat, ryes, sorghum, oats, rice, and corn. Large infestations of chinch bugs can destroy whole fields of grass crops. Some species of chinch bugs also feed on lawns, although they usually don't cause serious damage. There are usually two generations a year. Saltmarsh chinch bugs overwinter as adults in clumps of grass or fallen leaves and emerge in early spring. Females usually lay several hundred eggs.

BOXELDER BUG

Latin name: *Boisea trivittata*

Family: Rhopalidae

Identification: 0.5 inch long, black with red or orange fringes on forewings, red markings on thorax and head

Distribution: Eastern North America, west to Montana, Arizona, and Alberta

Comments: Boxelder bugs are common throughout North America wherever there are boxelder, maple, or ash trees, as they feed on the seeds of these trees. Although they usually feed on the seeds, they will sometimes feed on the leaves. They are most often found on female boxelder trees. The family that they belong to, Rhopalidae, are called scentless stink bugs. However, boxelder bugs produce a very strong pungent scent, and they often gather together in large groups because the secretion of their scent glands discourages predators. Although they are harmless, they can be pests when the temperature drops in the fall as they enter homes and other buildings to escape the cold.

ASSASSIN BUG

Latin name: *Zelus luridus*

Family: Reduviidae

Identification: 0.2 to 0.3 inch long, green, narrow elongated head

Distribution: Throughout the United States and southern Canada

Comments: Assassin bugs are especially fond of aphids. The sticky hairs on their powerful front legs hold their prey as it is consumed. After mating, females lay egg masses on leaves. Males protect the eggs until the bright green nymphs hatch. Nymphs often pile bits of leaves and debris on their head, thorax, and abdomen for camouflage.

Assassin bug nymphs often cover themselves with pollen and debris for camouflage.

Because assassin bugs feed on aphids and other pests, they are one of about fifty insect species that are raised commercially and sold to organic farmers and gardeners for controlling insect pests. One of the problems with using insects to control pests is that they often fly off. However, like all nymphs, assassin bug nymphs are wingless and also will feed on small insects. Although adult assassin bugs can fly, they usually stay put if the little insects on which they feed are reasonably plentiful. As organic farming and gardening become more widespread, commercial rearing of beneficial insects is likely to become an increasingly important industry.

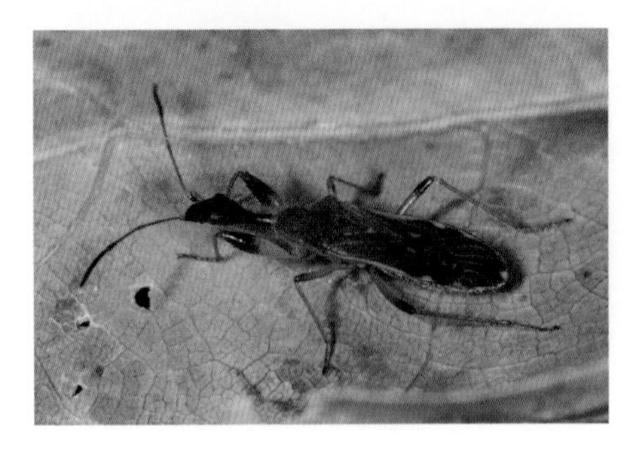

LONG-NECKED SEED BUG

Latin name: *Myodocha serripes*

Family: Rhyparochromidae

Identification: 0.4 to 0.6 inch long, brown, black head with long neck, enlarged black and yellow femora on front legs

Distribution: Eastern United States, west to North Dakota and Texas, southeastern Canada

Comments: Seed bugs were once considered one family, Lygaeidae, but now the family has been split into ten different families. Those in the family Rhyparochromidae are often referred to as dirt seed bugs because of their earthen-like coloration. The enlarged front legs of the long-necked seed bug are similar to those of assassin bugs. However, unlike assassin bugs, they are not employed to hold prey, as they feed on seeds, usually the seeds of strawberries and Saint-John's-wort. Long-necked seed bugs are sometimes in large enough numbers to cause serious damage to strawberries. They overwinter as adults.

GREEN STINK BUG

Latin name: *Acrosternum hilare*

Family: Pentatomidae

Identification: 0.5 to 0.7 inch long, shield-shaped, bright green with yellow margin on pronotum and head

Distribution: Throughout North America

Comments: Green stink bugs are found in gardens, woodlands, and crops across North America. Both adults and nymphs feed on plants. Adults have a preference for seeds and are pests of cotton, peas, beans, tomatoes, corn, soybeans, and eggplants. However, when seeds are not plentiful, green stink bugs will feed on leaves and cause damage to cherry, apple, orange, and peach trees. For many years a number of insecticides have been used to control these agricultural pests, and over time green stink bugs have developed resistance to most of them. There is a similar species in the southern states, the southern green stink bug (*Nezara viridula*), which is an imported species. It is also a pest of the same crops as the green stink bug. The green coloration is not the best way to distinguish the two species of green stink bugs, because their color can change as they age as well as being affected by their diet.

BROWN STINK BUG

Latin name: *Euschistus servus*

Family: Pentatomidae

Identification: 0.5 to 0.7 inch long, shield-shaped, brown, dark punctures on back

Distribution: Throughout North America

Comments: Brown stink bugs feed on the seeds of grains, fruits, and nuts. Like other stink bugs, both adults and nymphs have glands underneath their thorax from which they can secrete foul-smelling substances to discourage predators. The cocktail of chemicals includes aldehydes that have a pungent odor and cyanide compounds that give the secretions a rancid smell. Like other stink bugs, female brown stink bugs secrete a pheromone to attract mates. These pheromones can be synthesized. Farmers spray the man-made pheromone on the outskirts of their crops to attract stink bugs away from their fields.

TWO-SPOTTED STINK BUG, PREDATORY STINK BUG

Latin name: *Perillus splendidus*

Family: Pentatomidae

Identification: 0.3 to 0.5 inch long, flat, black with distinctive orange-red markings on pronotum and abdomen

Distribution: Southeastern Canada and eastern United States

Comments: Most adult stink bugs are camouflaged to blend in with the plants on which they live. However, this predatory stink bug has bright orange-red and black warning colors. Although many kinds of stink bugs are pests, because of their smell, they have gotten a somewhat undeserved bad reputation. However, stink bugs only release their foul-smelling defense toxins when they are threatened. Unless you put your nose right next to them, you usually can't smell the toxins. Some stink bugs, such as *Perillus splendidus,* feed on insects, including other stink bugs. If you see a black and orange-red stink bug in your garden, leave it alone.

SHIELD BUGS, ROUGH STINK BUGS

Latin name: *Brochymena* sp.

Family: Pentatomidae

Identification: 0.5 to 0.8 inch long, rough, gray, prominent thoracic sclerite

Distribution: Southern Canada and throughout most of the United States

Comments: Rough stink bugs are found in orchards, fields, and woodlands, where they feed on the sap from twigs of many kinds of bushes and trees including pine, willow, sumac, elm, cherry, ash, and apple. Their rough appearance and dull gray color blends in with the trees on which they feed. Although they usually feed on tree sap, from time to time they will feed on insects. This is characteristic of a number of species of stink bugs.

Female jagged ambush bug

JAGGED AMBUSH BUG

Latin name: *Phymata americana*

Family: Reduviidae

Identification: 0.2 to 0.5 inch long, heavily armored, prominent front legs, broad-shouldered

Distribution: Throughout the United States and southern Canada

Comments: Ambush bugs are particularly formidable predators that will take on insects that are larger than themselves, or even spiders. They have an especially hard exoskeleton and strong front raptor-like legs that they employ as a praying mantis would to grasp and hold their prey. After securing their prey insect, they stab it with their stylets, inject poisons to immobilize it, and suck out the liquefied tissue. On sunny days, these stalky little bugs can be found on flowers, patiently waiting for flies, bees, and other insects to come by. They are camouflaged to blend in with flowers, especially yellow ones. Jagged ambush bugs overwinter as adults.

Male jagged ambush bug

CREEPING WATER BUGS, SAUCER BUGS

Latin name: *Pelocoris* sp.

Family: Naucoridae

Identification: 0.2 to 0.4 inch long, oval-shaped, pointed abdomen, hook-shaped grasping front legs

Distribution: Eastern United States, west to the Mississippi River

Comments: Creeping water bugs are well-camouflaged predatory hemipterans that live in freshwater ponds, rivers, and streams. They have powerful, hairy hind legs that enable them to swim quite fast. They also fly very well, although they seldom take to the air. When they come to the surface to breathe, they store a bubble of air underneath their wings. Creeping water bugs often sit on the bottom of the pond or under rocks, where they are difficult to see. A good way to find them is to uproot an aquatic plant, remove the mud from around the plant root, and search through the mud. *Pelocoris* is the common genus in eastern North America. The genus *Ambrysus* is found in western states.

WATER SCORPIONS, NEEDLE BUGS, WATER STICK INSECTS

Latin name: *Ranatra* sp.

Family: Nepidae

Identification: 0.6 to 2 inches long, brown, stick-like, superficial resemblance to scorpions

Distribution: Throughout the United States and southern Canada

Comments: Water scorpions get their name because they look a bit like scorpions. However, they look more like twigs, which acts as their camouflage. They live in quiet ponds among the aquatic plants, where they are very difficult to see as they crawl slowly among the plants. They breathe air from the surface through two long tubes that are attached into a single long tube, which extends from their abdomen. Water scorpions are stealthy ambush predators. They cling to vegetation with their four hind legs and slowly move their raptor-like front legs out when they see an insect, small fish, or pollywog. When the prey gets close enough, they reach out, grab it, and hold it in a jackknife grip.

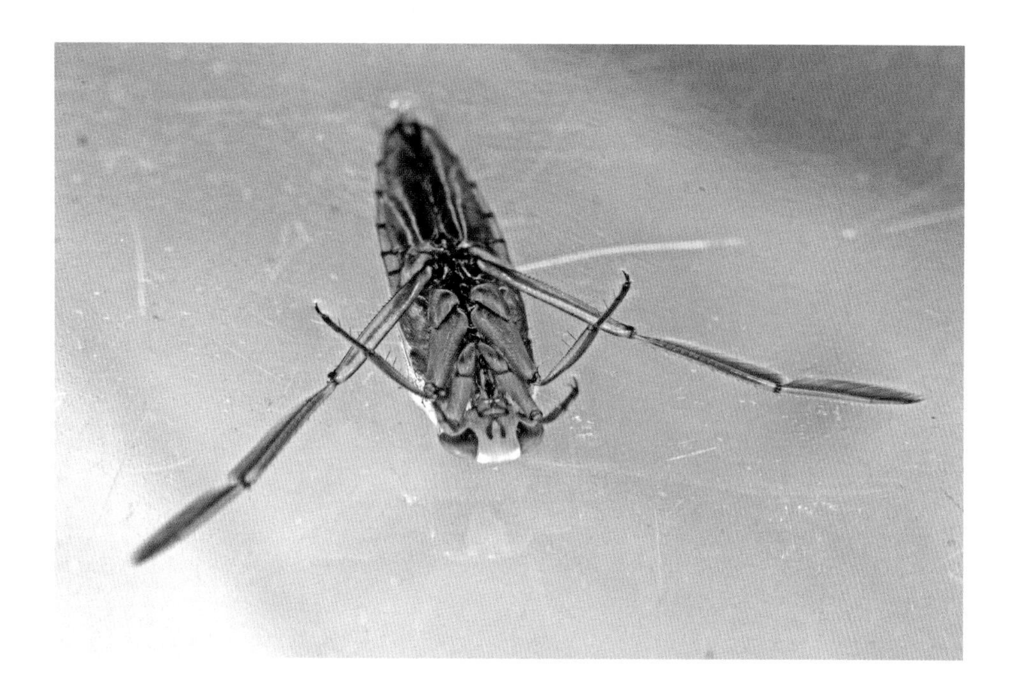

BACKSWIMMERS, WATERBEES

Latin name: *Notonecta* sp.

Family: Notonectidae

Identification: 0.5 to 0.7 inch long, swim upside down, shiny white back, large red eyes

Distribution: Throughout the United States and southern Canada

Comments: Backswimmers swim and rest on their backs. Since their eyes are mostly below the water line, they can see what goes on in the water below them. They have a shiny white back that blends in with the light sky so that it is less likely that underwater predators will see them. Because they are lighter in weight than the water, backswimmers can float without consuming energy. When they dive, they carry a bubble of air under their wings. Oxygen from the bubble diffuses through the bug's spiracles.

Backswimmers prey on insects, small tadpoles, and tiny fish. They fly very well and often take flights in the fall in order to find new ponds. However, they also find swimming pools, where they can be a nuisance. If you see them in a swimming pool, be careful not to handle them too roughly because they can bite. The bite can be as bad as a bee sting, which is why they are sometimes called "waterbees."

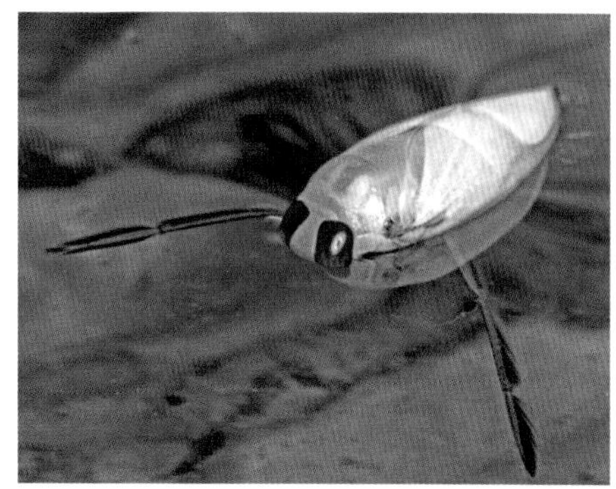

I flipped this backswimmer over. The underside of backswimmers is white to blend in with the sky.

WATER BOATMEN, WATER CICADAS

Latin name: *Hesperocorixa* sp.

Family: Corixidae

Identification: 0.2 to 0.6 inch long, oval flattened body, fine lines on forewings, long hairy oar-shaped rear legs, short front legs

Distribution: Throughout the United States and southern Canada

Comments: Unlike most other aquatic hemipterans, water boatmen are herbivorous, feeding on aquatic plants and algae. In order to breathe underwater, these bugs obtain bubbles from the surface and hold them under their wings. Water boatmen frequently fly and are attracted to lights. To take off from the water, they hurtle themselves through the surface of the water, open their wings, and are airborne. During mating season, water boatmen chirp, which is why they are sometimes called "water cicadas." Females attach their eggs to plants. Some native Mexicans collect water boatman eggs from plants, dry them, and grind them into flour.

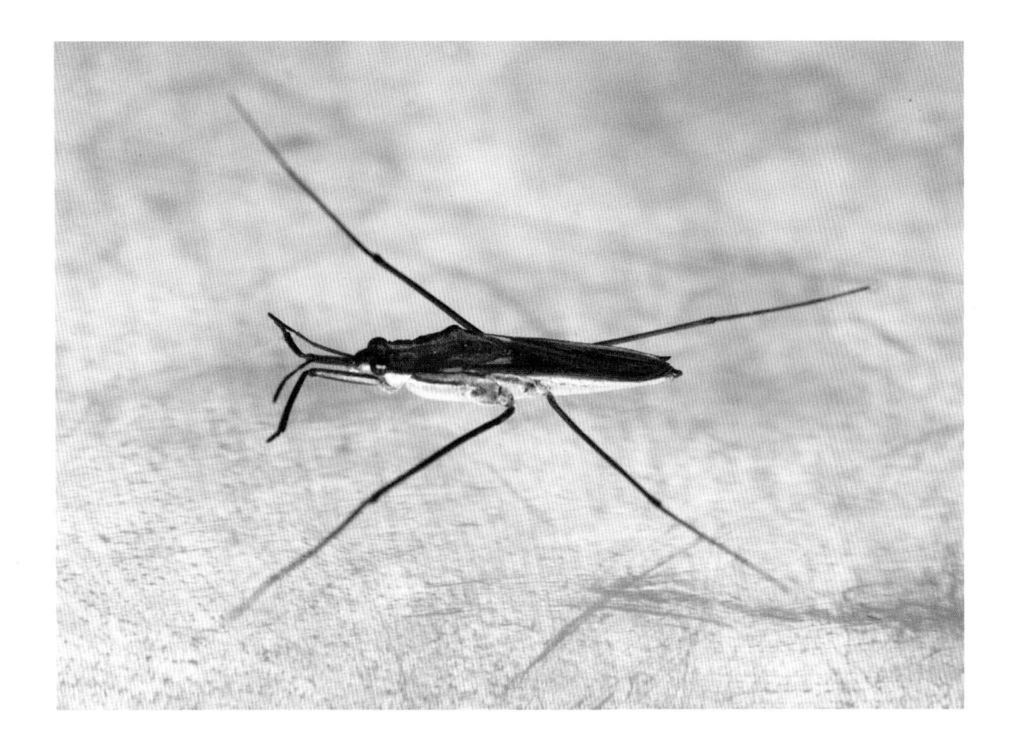

WATER STRIDERS, JESUS BUGS

Latin name: *Gerris* sp.

Family: Gerridae

Identification: 0.5 to 0.7 inch long, black or dark-colored, slender, long middle and back legs, raptor-like front legs

Distribution: Throughout North America

Comments: Water striders live in ponds, lakes, and rivers, where they skate across the water in rapid bursts of speed. Several anatomical features contribute to the ability of these insects to walk on water. The legs of water striders are strong and long, which distributes their weight over a large area. Numerous microscopic non-wetting hairs cover the surface of the entire insect, preventing water from waves or rain from sticking to their bodies. Tiny hairs on their legs provide a hydrophobic surface and increase the surface area of the rear four legs. Water striders use their middle legs for propulsion and their rear legs for steering. Their front legs touch the surface of the water to detect ripples created by insects that fall into the water, on which they feed. The front legs are also employed to grab and hold these insects. Some individuals have wings, but others are wingless.

WATER STRIDERS

Latin name: *Trepobates* sp.

Family: Gerridae

Identification: 0.2 to 0.3 inch long, short abdomen

Distribution: Throughout eastern North America, less common than the genus *Gerris*

Comments: *Trepobates* species are small water striders with a long thorax and short abdomen. They are found skating in groups on quiet lakes. Setae on their front legs can detect ripples on the water surface caused by the struggles of little insects and spiders that fall into the water. These setae can also detect danger. When approached, they skate rapidly away, which makes them very difficult to catch. A genus of water striders that few have seen is *Halobates,* as they live far out at sea. Most species are tropical, but one species lives off the coast of California. Although some kinds of insects live in tidal pools and salt marshes, *Halobates* is the only genus of truly marine insects.

GIANT WATER BUGS, TOE-BITERS

Latin name: *Lethocerus* sp.

Family: Belostomatidae

Identification: 0.7 to 1 inch long, large flattened oval body, short breathing tube, raptor-like front legs

Distribution: Throughout North America

Comments: Giant water bugs are large aquatic hemipterans. They often hold on to aquatic plants and wait for insects, snails, small fish, amphibians, and crustaceans that they feed upon to pass by. When they fly, giant water bugs breathe through spiracles, but when they are in the water, they breathe through a respiratory tube that protrudes from their abdomen like a snorkel.

Giant water bugs are one of the relatively few nonsocial insects that exhibit parental care. In many species, females lay their eggs in the mud and their mates guard them. In other species, females lay their eggs on the back of their mate. Males are not very happy about this arrangement and often try to fight off their mate. However, female giant water bugs are larger than males and eventually succeed in laying their eggs on his back. Once the eggs are there, the males care for them until they hatch and the first instar naiads swim off.

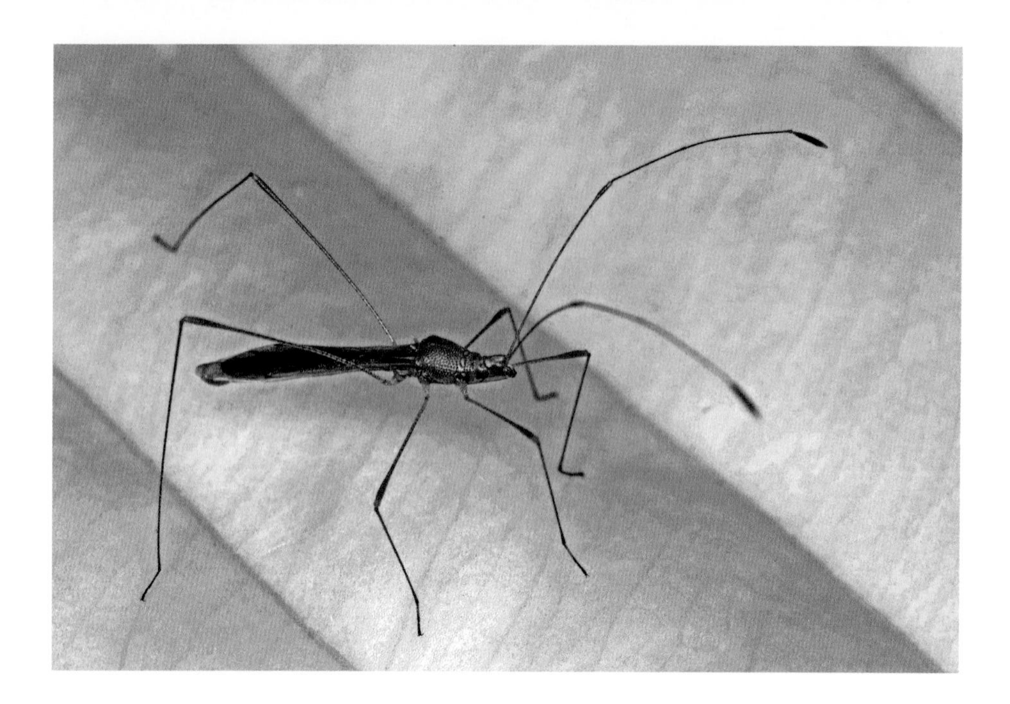

SPINED STILT BUG

Latin name: *Jalysus wickhami*

Family: Berytidae

Identification: 0.2 to 0.3 inch long, cigar-shaped, reddish brown, exceptionally long thin legs

Distribution: Throughout North America

Comments: Although they can fly, stilt bugs usually use their exceptionally long, thin legs to walk on the stems and leaves of plants. They feed on the stems of many kinds of plants, including tomato, soybean, tobacco, alfalfa, and cotton, and can sometimes be pests of tomato plants. If a spined stilt bug happens to come across an aphid, small caterpillar, or other small soft-bodied insects on its journey, it devours the little insect. This is beneficial for the spined stilt bug because they will live longer and produce more eggs if they supplement their plant diet with insects. In fact, spined stilt bugs are often reared by tobacco farmers because they feed on the eggs and caterpillars of tobacco budworm (*Heliothis virescens*) and tobacco hornworm (*Manduca sexta*). Spined stilt bugs overwinter as adults. Their nymphs have five instars and require both plant sap and insects to develop.

BLACK DAMSEL BUGS

Latin name: *Nabis* sp.

Family: Nabidae

Identification: 0.1 to 0.2 inch long, black with prominent rostrum, short wings or wingless

Distribution: Throughout North America

Comments: Damsel bugs are predacious insects that capture small insects, especially aphids, but they also prey on leafhoppers, sawfly larvae, mites, and beetle eggs. If none of these are available, damsel bugs are happy to eat each other. They hold their prey in their raptor-like forelegs as a praying mantis would. Black damsel bugs are found in fields, where they prowl grasses looking for the little insects and insect eggs on which they feed. They are also found in gardens and crops, especially soybeans and alfalfa. Needless to say, gardeners and farmers are happy to see them.

Broad-headed bug nymphs are ant mimics.

BROAD-HEADED BUGS

Latin name: *Alydus* sp.

Family: Alydidae

Identification: 0.5 to 0.8 inch long, broad head, black with long segmented antennae

Distribution: Throughout North America

Comments: Broad-headed bugs and their nymphs feed on the seeds of trees and many kinds of plants, including cultivated plants. Although they prefer to feed on seeds, they also feed on stems and foliage when seeds are scarce. They sometimes cause significant damage to gardens and farms. Young fruit becomes deformed when they develop after they have been attacked by broad-headed bugs, and seeds become flattened and shriveled. Broad-headed bugs are wasp mimics, especially when they fly, and *Alydus* nymphs are very convincing ant mimics. They even walk like ants. When I first saw one of these nymphs, I would have surely thought it was an ant if I had not previously seen pictures of broad-headed bug nymphs.

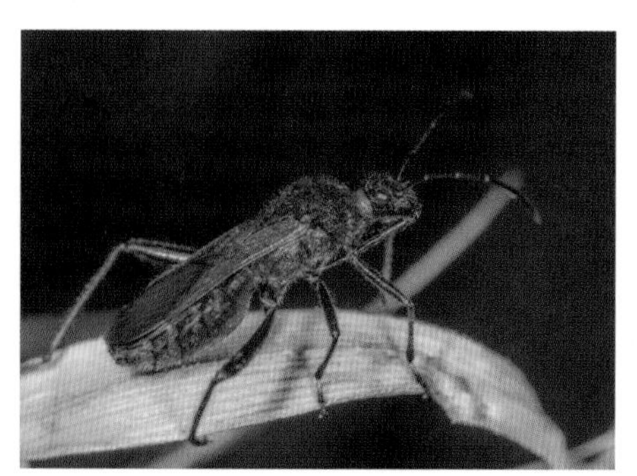

When in flight, broad-headed bug nymphs mimic wasps.

BROAD-HEADED BUG, LUPINE BUG

Latin name: *Megalotomus quinquespinosus*

Family: Alydidae

Identification: 0.5 to 0.8 inch long, slender, brown, broad head, long segmented antennae with white markings

Distribution: Throughout the United States and southern Canada

Comments: These broad-headed bugs are sometimes called lupine bugs because they feed on the seeds of the beautiful native North American perennial flowering plants known as lupine. In addition, they will feed on the seeds of many other kinds of plants and trees. They are active, alert, and fly well. When they are threatened, broad-headed bugs secrete noxious chemicals called allomones (butyric and hexanoic acids) from scent glands in their thorax. They look very much like wasps in flight, which helps protect them from birds. Like some other broad-headed bugs, their nymphs are ant mimics.

A broad-headed bug nymph mimicking an ant

WESTERN CONIFER SEED BUG

Latin name: *Leptoglossus occidentalis*

Family: Coreidae

Identification: 0.6 to 0.8 inch long, orange and brown with thin white lines in the shape of a rectangle on forewings

Distribution: Throughout the United States and southern Canada

Comments: Until the 1950s, this species of leaf-footed bug was limited to western states, but it can now be found across the continent. Western conifer seed bugs feed on the sap of developing cones of a number of kinds of pine trees. When they are present in large numbers, they can seriously harm pine trees, especially young trees. These seed bugs overwinter as adults and frequently enter homes, where they can be a nuisance, especially because they emit a buzzing noise when they fly as well as release foul-smelling chemicals when they feel threatened. They can occasionally be a more serious problem by puncturing polyethylene tubing that is used in plumbing. Unlike their relatives that feed on insects, they do not produce toxins. Also, they cannot inflict a painful bite.

DISTINCT LEAF-FOOTED BUG

Latin name: *Merocoris distinctus*

Family: Coreidae

Identification: 0.3 to 0.5 inch long, brown, humpback, large flat hind tibia, 2 prominent ocelli, 4-segment antennae

Distribution: Northeastern United States and southeastern Canada

Comments: Leaf-footed bugs can be identified by their large, usually flattened hind legs that are reminiscent of a leaf. They are sometimes called twig wilters or tip wilters because they inject enzymes that soften the young twigs that they feed on, which can cause them to wilt. Like other leaf-footed bugs, distinct leaf-footed bugs have stink glands that emit a foul odor when they are attacked or handled. Distinct leaf-footed bugs overwinter as adults in protected areas such as in woodpiles, under peeling bark, or in cracks in trees, often in groups of dozens of individuals.

LEAF-FOOTED BUGS

Latin name: *Leptoglossus* sp.

Family: Coreidae

Identification: 0.3 to 0.6 inch long, brown, leaflike enlargements of hind tibia, white markings on forewings

Distribution: Throughout North America

Comments: Species of the genus *Leptoglossus* are pests of many crops, including fruits, vegetables, grains, nuts, and ornamentals. They usually are not in high enough numbers to be considered major pests, but even when there are relatively few bugs, they can be a problem with tomatoes, where insect feeding can cause depressions and discoloration. Although species of *Leptoglossus* are usually not common enough to cause serious damage to most crops, occasionally large infestations occur that can destroy sizable proportions of a crop. Both nymphs and adults have stink glands that can release foul-smelling chemicals when the insects sense danger.

BED BUG

Latin name: *Cimex lectularius*

Family: Cimicidae

Identification: 0.15 to 0.2 inch long, 0.06 to 0.19 inch wide, light brown to reddish brown, flat oval body, abdomen with banded appearance, forewings greatly reduced, hindwings absent

Distribution: Throughout North America

Comments: Bedbugs are primarily nocturnal insects that hide during the day and feed on human blood at night. Generally, their bite is not painful, so sleeping people do not know that they have been bitten. Bedbugs were uncommon in the 1940s, but since then, they have had a resurgence. A number of hypotheses have been put forth for the increase in bedbug populations in North America. Probably the most plausible is that they have developed resistance to the insecticides that are employed to kill them. Another idea is that people formerly employed insecticides to control cockroaches but now use cockroach traps, which use a pheromone to attract the cockroaches and a sticky substance to attach them to the trap. Bed bugs are killed by the insecticide but are not attracted to cockroach traps. Another idea is that the increase in travel has caused the spread in bedbugs because they can enter clothing and suitcases.

FLAT BUGS, FLAT BARK BUGS, BROWN MARMORATED STINK BUGS

Latin name: *Aradus* sp.

Family: Aradidae

Identification: 0.2 to 0.4 inch long, very flat, dark, abdomen extends beyond wings

Distribution: Throughout the United States and southern Canada

Comments: Flat bugs usually live under bark. Adults and nymphs are usually found in dead and dying trees because the space between the bark and the wood of these trees is home to many kinds of fungi on which flat bugs and their nymphs feed. They also can be found in piles of old logs where there are lots of fungi. Flat bugs have unusual mouthparts that are coiled like a spring. Nymphs, which are usually more numerous than adults, have very distinct abdominal segments. There are about seventy-five species of *Aradus* in North America. Identifying them to species is a task for an expert on this genus. One southern species, *Aradus gracilis,* lives in trees that have been burnt.

A species of Asian lanternfly, *Pyrops viridirostris*, which has the typical lantern-like appearance

SPOTTED LANTERNFLY

Latin name: *Lycorma delicatula*

Family: Fulgoridae

Identification: 0.8 to 1.1 inches long, black head, gray wings with black spots, yellow abdomen with black and white bands

Distribution: Throughout Pennsylvania, with a few in New York, Delaware, and Virginia

Comments: Until 2014, lanternflies were unknown in North America, but in 2014, spotted lanternflies were found in Pennsylvania and have been spreading to neighboring states. Lanternflies are named for their long proboscis that gives them a shape reminiscent of an old-fashioned lantern. There is great concern about the spotted lanternflies because they feed on many plants. This Asian species was accidentally introduced into South Korea in 2006 and has expanded its range and now feeds on some sixty-five plant species. As these hemipterans damage agricultural products in South Korea, authorities are very concerned that they will expand their range and cause more damage. There are bans on moving firewood, lawnmowers, and other things that might contain the insects or their eggs from certain Pennsylvania counties. Whether these measures will contain the spotted lanternfly or whether this hemipteran will become a major threat to agriculture remains to be seen.

21 BEETLES

Beetles are by far the largest order of insects in North America and the world. It is estimated that one in every four species of animals in the world is a beetle. About 30,000 species have been described in North America, ranging in size from that of a grain of sand to more than 2 inches in length. All beetles have chewing mouthparts, with well-developed hard mandibles that are often used for crushing seeds or gnawing wood. The forewings of most beetles are hardened into stiff plates called elytra, which lie over and protect their hindwings and abdomen. The hindwings of beetles are delicate and usually larger than the forewings and folded such that they can fit under the elytra and open when the elytra are lifted in preparation for flight.

Not all species of beetles can fly. With the exception of tiger beetles, jewel beetles, and a few others, beetles fly awkwardly and steer poorly. When beetles fly, their elytra extend outward and are employed for lift and steering, somewhat like the wings of an airplane, while the hindwings are used for propulsion.

ACORN WEEVILS

Latin name: *Curculio* sp.

Family: Curculionidae

Identification: 0.2 to 0.4 inch long, with long curved snout

Distribution: Common throughout North America

Comments: These little weevils employ their snout to chew holes in acorns, beechnuts, hazelnuts, or hickory nuts. After feeding on the meat (endosperm) of the nut, females turn around and use their long ovipositor to lay an egg or eggs in the nut, where the larvae will mature as they feed on the endosperm, protected by the nutshell. When the nuts fall to the ground in the fall, larvae chew holes in the nutshell and burrow into the ground, where they overwinter, pupate, and emerge the following year or after two years as adult weevils. For the reader who is eager to see or own one of these little weevils, simply collect some acorns from the ground in the fall and place them in a pot of water. The ones that float may contain one or more weevil larvae. Place these acorns in a flowerpot with a few inches of soil, cover the flowerpot with netting, and put it outside. After a year or maybe two, you may find some acorn weevils.

The hindwings of most beetles are folded under the elytra. They unfold when the beetle lifts its elytra.

FULLER ROSE BEETLE, FULLER ROSE WEEVIL

Latin name: *Naupactus cervinus*

Family: Curculionidae

Identification: 0.25 to 0.3 inch long, dark brown with pinkish scales

Distribution: Widespread in North America

Comments: Fuller rose beetles are small flightless beetles. Like most other flightless beetles, their elytra are fused. For a beetle that cannot fly, Fuller rose beetles get around pretty well. The species was introduced to California in 1896, probably from Mexico, and now can be found coast to coast and in much of the rest of the world. In North America, they are more common in the eastern and central regions of the continent.

Fuller rose beetles are all females that reproduce by parthenogenesis. They feed on a variety of plants including peach trees, strawberries, rose bushes, and potato plants, but they do the most damage in citrus groves. When these nocturnal beetles feed, they chew notches in leaves and damage the protective coat of flower buds. Larvae feed on roots, which can seriously weaken trees when the larvae are numerous.

APPLE CURCULIO (WEEVIL)

Latin name: *Anthonomus quadrigibbus*

Family: Curculionidae

Identification: 0.2 to 0.3 inch long, elytra with ridges and small patches of orange

Distribution: Throughout eastern North America

Comments: *Curculio* is the Latin word for "weevil." Apple curculios feed on a number of fruit trees, including crab apple, juneberry, and hawthorn, but do the most damage to apple orchards. After feeding on apples, female weevils turn around, poke their ovipositor in the hole that they made while feeding, and lay their eggs in the apple. The apples that are attacked are misshapen and lumpy and have brown or soft areas. Cutting the apples open reveals both weevils and their larvae. The apples drop from the tree in the fall, and adult weevils crawl from the apple, overwinter in the soil, and return to the trees in the spring. The holes that these weevils make can also be entryways for other insects that feed on apples. Only some varieties of apples are attacked by apple curculios. Delicious apples are especially vulnerable. Apple curculios can be controlled, but not eliminated, by cleaning up the apples that have fallen from the trees.

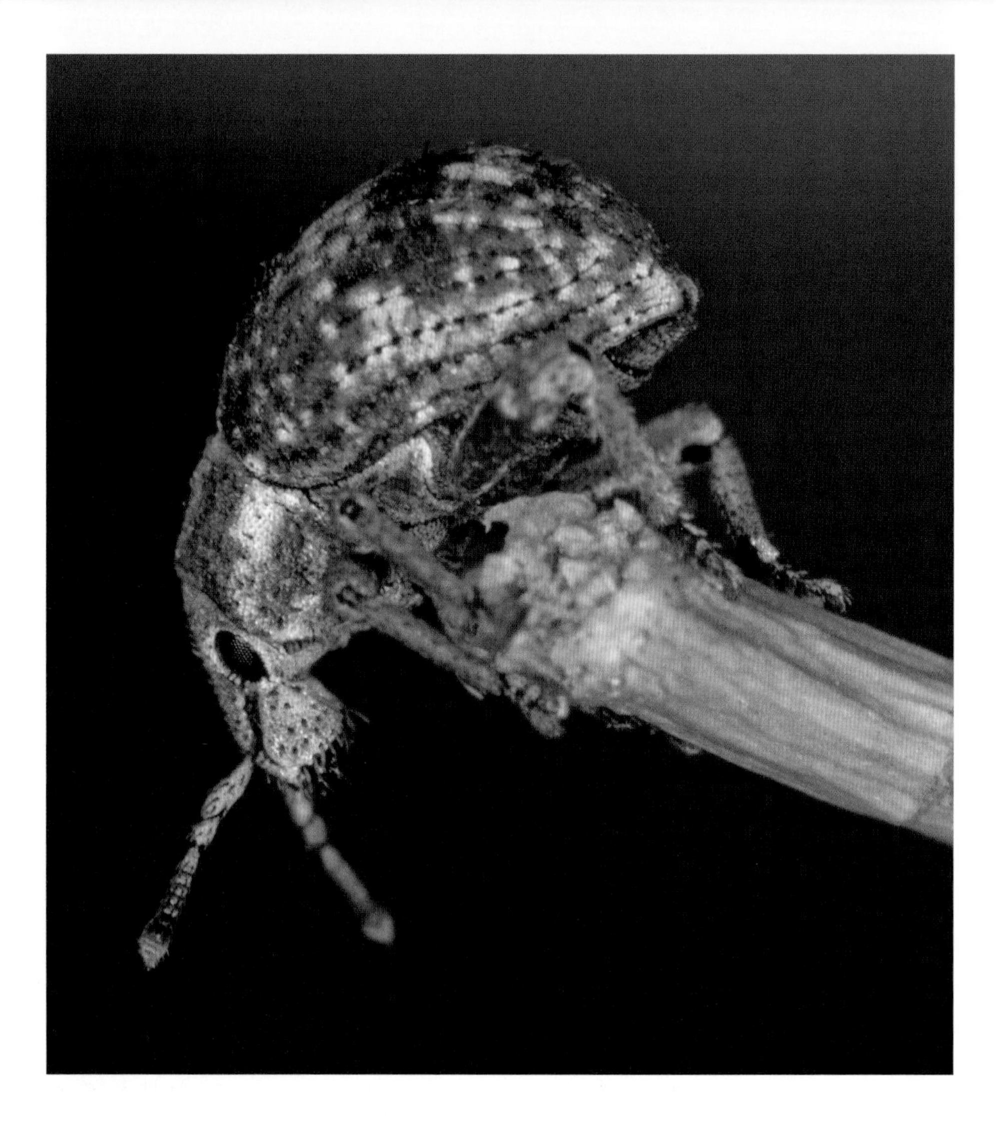

IMPORTED LONG-HORNED BEETLE

Latin name: *Calomycterus setarius*

Family: Curculionidae

Identification: 0.1 to 0.2 inch long, oval, dark reddish brown with whitish scales, broad nose

Distribution: Ontario, eastern Quebec, eastern United States to Iowa and Kansas

Comments: These little flightless weevils feed on a variety of plants. Their larvae live in soil, where they feed on the roots of many different plants including aster, turf grass, and clover. Imported long-horned beetles and their larvae don't cause many problems, although they can be a nuisance when they enter homes and buildings. These beetles are native to Japan and were first reported in the United States in 1929. They are all parthenogenetic females.

NO COMMON NAME

Latin name: *Polydrusus sericeus*

Family: Curculionidae

Identification: 0.2 to 0.3 inch long, covered with iridescent light green scales, long antennae, elytra without setae

Distribution: Prince Edward Island, south to Connecticut, west to Wisconsin and Illinois

Comments: These pretty little green broad-nosed weevils were imported to North America from Eurasia. Like other broad-nosed weevils, *Polydrusus sericeus* cannot fly. Adults feed on the leaves and young blossoms of hardwood trees, especially birch, basswood, ironwood, maple, and fruit trees. They can cause serious damage to apple, pear, and cherry trees by feeding on buds, blossoms, and shoots. Adult females lay their eggs in the summer or fall in cracks in bark or on leaves of their host plant. When the larvae hatch, they crawl or fall to the ground, where they feed on roots of the host plant. Larvae overwinter and pupate in the spring.

BARK BEETLE

Latin name: *Pityokteines sparsus*

Family: Curculionidae

Identification: 0.07 to 0.1 inch long, dark brown, cylindrical with dense setae

Distribution: Newfoundland to Alberta, northern United States

Comments: These tiny beetles chew tunnels in the inner bark of trees. In the case of this bark beetle, the trees are balsam fir (*Abies balsamea*) and other conifers. Other species of bark beetles attack hardwood trees. After mating, females lay their eggs in tunnels that they chew in the inner bark. When the larvae hatch, they dig separate tunnels off of the original tunnel that their mother made, which creates a "gallery" of tunnels. Some bark beetles attack dead or dying trees, which aid in the tree's decomposition. However, the species that attack live trees can cause tremendous damage. Healthy trees may produce insecticides, as well as sap and resin that kills, injures, or traps bark beetles. However, sometimes there are so many beetles that they overwhelm the trees' defenses, ultimately killing them. This can have a devastating effect on lumber, wildlife, and property values.

ROSE CURCULIO

Latin name: *Merhynchites bicolor*

Family: Attelabidae

Identification: 0.2 inch long, red, black legs and antennae

Distribution: Ontario south to South Carolina, west to North Dakota and Kansas

Comments: Rose curculios feed on roses as well as raspberries and blackberries. Females employ their long snout to bore a hole deep into rosebuds. After feeding on the tissue, they turn around and lay their eggs in the bud. When the eggs hatch, the larvae feed on the flower petals. This kills some flowers, and the petals on those that open are riddled with little holes that were made by the adult beetle. Damage is also caused by the larvae, which feed on the buds. This usually weakens the buds so that they break off and fall to the ground. When the larvae are ready to pupate, they crawl from these fallen buds and pupate in the ground. If there are very few buds on the rose bush, rose curculios either feed on the tips of the rose shoots, which results in the death of the terminals, or feed on the stem of the buds, which causes the bud to wilt and die.

LEAF-ROLLING BEETLE

Latin name: *Homoeolabus analis*

Family: Attelabidae

Identification: 0.2 to 0.25 inch long, red elytra and pronotum, black legs and antennae

Distribution: Ontario to Florida, west to Manitoba, Texas, and Kansas

Comments: Leaf-rolling beetles are primitive weevils. They can be distinguished from the more derivative (modern) weevils because their antennae are not elbow-shaped. Inseminated females use their mandibles to chew the base of leaves and cut pieces into rectangular shapes. Cutting the leaves softens them so they are easier to roll. The beetle then lays an egg on the cut part of the leaf and uses her forelegs to roll it around her egg. She may make a dozen rolls, which looks like a tiny rolled-up rug. When the egg hatches, the rolled leaf serves as a shelter and camouflage for the larva as it feeds on the leaf. Each species of leaf-rolling beetle rolls leaves in a characteristic manner.

FUNGUS WEEVIL

Latin name: *Euparius paganus*

Family: Anthribidae

Identification: 0.15 to 0.25 inch long, short snout, brown mixed with yellow head with dense pubescence, antennae not elbowed, elytra pale gray or whitish scales with black marks

Distribution: Quebec, south to Florida, west to Manitoba, Montana, Nebraska, and Texas

Comments: Fungus weevils are unusual in that they do not have elbowed antennae like most weevils. Both adults and larvae feed on fungi. Adults graze on the surface of fungi, whereas larvae bury into the fungi. Although these weevils do not cause problems, some beetles of the family Anthribidae do. The coffee bean weevil (*Araecerus fasciculatus*) is a cosmopolitan species that, despite its name, is a serious pest of many stored grains. A predatory European species, *Anthribus nebulosus*, is now established in the northern United States, where it was imported to control scale insects.

Asian multicolored lady beetles mating. The number of spots is variable.

ASIAN MULTICOLORED LADY BEETLE, HARLEQUIN LADY BEETLE, ASIAN LADY BEETLE

Latin name: *Harmonia axyridis*

Family: Coccinellidae

Identification: 0.25 inch long, black M design in the center of pronotum, variable number of black spots on elytra

Distribution: Throughout southern Canada and the United States

Comments: An outstanding feature of these beetles is the variation in the pattern of black spots on their red or orange elytra. Some individuals have no spots, while others can have up to twenty. Adult beetles and larvae feed on small soft-bodied insects, especially aphids. Because they are particularly effective in controlling aphids, this species has frequently been imported to North America. In some places, they have displaced the native species by reducing the population of aphids and by feeding on the eggs and larvae of the native lady beetle species.

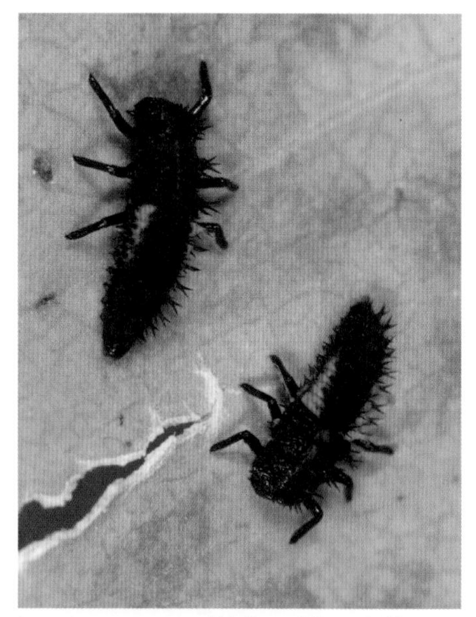

Larvae consume more aphids than adults, probably because they are growing.

The yellow, orange, and red warning colors indicate that these and other lady beetles are poisonous. In this case, the poison is in the beetle's hemolymph and is secreted through leg joints in an action called "reflex bleeding." They can be a nuisance when they congregate in large numbers in homes and buildings and release a foul-smelling substance.

These lady beetles have several different patterns. This one looks like a smiley face.

FOURTEEN-SPOTTED LADY BEETLE, CREAM-SPOT LADYBIRD, POLKADOT LADYBIRD

Latin name: *Propylea quatuordecimpunctata*

Family: Coccinellidae

Identification: 0.16 to 0.2 inch long, various colors and patterns

Distribution: Throughout North America

Comments: This lady bug comes in a variety of colors: Some are brown or yellow with white spots; others are yellow, but instead of spots, they have a pattern that looks like a smiley face.

The fourteen-spotted lady beetle is thought to have originated in Japan. It was imported to Europe and from there to North America. Like other ladybird beetles, adults and larvae feed on aphids and other small soft-bodied insects and their eggs, including other lady bugs. The eggs of *Calvia quatuordecimguttata* are coated with an acid that protects them from being eaten. The yellow color reminds predators that they are poisonous. Like other lady bugs, they secrete toxic chemicals from their leg joints when they are attacked.

This is the most common spot pattern of the fourteen-spotted lady beetle.

HYPERASPIS LADY BEETLES, WHITE WAX LADIES

Latin name: *Hyperaspis* sp.

Family: Coccinellidae

Identification: 0.1 to 0.2 inch long, round body, usually black with one red spot on either side of elytra, white face

Distribution: Eastern United States and southern Canada

Comments: Hyperaspis lady bugs are small, roundish beetles. Their larvae, as well as those of some other genera of lady beetles, mimic the mealybugs that they feed on. The larvae produce wax from glands on their sides, which covers them and makes them look like mealybugs. This disguise is not intended to fool the mealybugs that they feed on since mealybugs can't run away or fight back. Rather, the disguise is meant to fool the ants that feed on the honeydew that the mealybugs produce and in turn protect the mealybugs. So next time you see a white waxy-looking bug, give it a poke. If it moves, it is a lady bug larva, and if it stays put, it is a mealybug.

Larvae of hyperaspis lady beetles cover themselves with wax in order to look like the mealy bugs that they feed on.

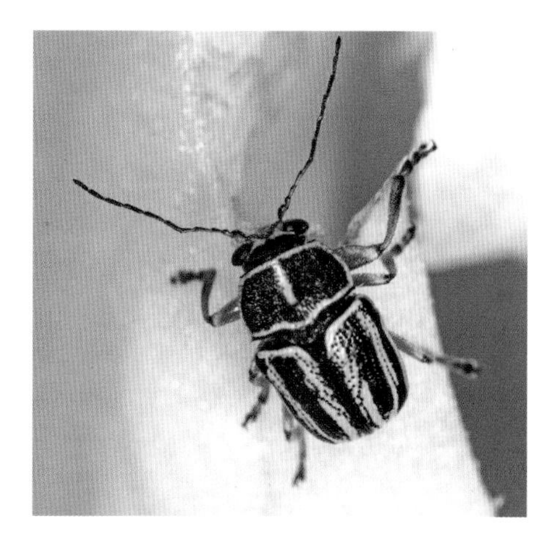

CASE-BEARING LEAF BEETLE

Latin name: *Pachybrachis othonus*

Family: Chrysomelidae

Identification: 0.04 to 0.6 inch long, variable shapes, usually colorful

Distribution: Throughout North America

Comments: About 2,000 species of leaf beetles have been described. Many are colorful, and they all feed on plants. Although some species are agricultural pests or control obnoxious weeds, the large majority live out their lives in meadows and woodlands munching on weeds or other plants without having a serious effect on our lives or even the plants on which they feed. Adults and larvae of most leaf beetle species feed on a particular species of plant or a few related species, but some are not so fussy about the plants on which they feed.

DOGBANE BEETLE, GOLDEN BEETLE

Latin name: *Chrysochus auratus*

Family: Chrysomelidae

Identification: 0.3 to 0.5 inch long, iridescent green head and thorax, iridescent red abdomen, color changes when viewed from different angles

Distribution: Maritime Provinces, west to Alberta, and all of continental United States except California and Florida

Comments: The surface of dogbane beetles is made of stacks of tiny slanting plates that overlay pigment. Light reflects from the surface of the plates and the pigment such that when the beetle is viewed at different angles, its color appears to changes. Dogbane beetles spend their entire lives on dogbane. Dogbane is a shrub that is similar to milkweed and, like milkweed, is poisonous. Insects that feed on these plants usually sequester the poison and display bright warning colors. Adult beetles feed on the leaves of the plant, while the larvae feed on the roots. Female beetles lay two or three eggs on the underside of each of a number of leaves and cover them with frass for protection. When the eggs hatch, the larvae chew through the frass and fall to the ground, where they feed on roots and eventually pupate in the ground.

FALSE POTATO BEETLE

Latin name: *Leptinotarsa juncta*

Family: Chrysomelidae

Identification: 0.3 to 0.4 inch long, yellow elytra with 3 black stripes, orange pronotum with black markings

Distribution: Pennsylvania, south to Florida, west to Texas

Comments: The false potato beetle (*Leptinotarsa juncta*) looks almost identical to the potato beetle (*Leptinotarsa decemlineata*), but it doesn't feed on the potato plant. The history of the potato beetle itself is an example of an insect's triumph over man. Potato beetles are present across most of North America, Europe, and Asia but are believed to have originated in Mexico, where they fed on Buffalo bur, a member of the potato family. When the beetles expanded their range into North America around 1840, they switched their preference to potato plants. Many insects are able to change their food plant when a closely related plant is more prevalent. Subsequently, potato beetles were accidentally imported to Europe, where they have been a major pest for many years. DDT was effective in controlling potato beetles, but when they became resistant, other insecticides were used. Over the years some populations of potato beetles have become resistant to all insecticides.

SKELETONIZING LEAF BEETLES

Latin name: *Trirhabda* sp.

Family: Chrysomelidae

Identification: 0.3 to 0.5 inch long, yellow with black stripes on elytra, 3 black spots on pronotum, 1 black spot on head

Distribution: Quebec to Florida, west to Texas

Comments: There are twenty-four species of the genus *Trirhabda* in North America. Most species feed on shrubs of the Asteraceae family, which includes dandelions, thistles, and other weeds, or the Hydrophyllaceae family, most of which are of little economic importance. As most entomologists study insects that impact people, these colorful little beetles have not been studied very extensively. Skeletonizing leaf beetles, as well as many other skeletonizing insects, are referred to as skeletonizers because their larvae feed on the soft part of leaves while leaving the veins intact. The leaf veins serve as a scaffolding for the larvae as they munch away at the tissue between them.

SPOTTED CUCUMBER BEETLE, SOUTHERN CORN ROOTWORM (LARVAE)

Latin name: *Diabrotica undecimpunctata*

Family: Chrysomelidae

Identification: 0.2 to 0.3 inch long, yellow elytra with 10 black spots, yellow or greenish pronotum

Distribution: Ontario and Quebec, widespread in the United States

Comments: Spotted cucumber beetles feed on the leaves of a number of crops. Their larvae, which are called southern corn rootworms, feed on the roots of plants, especially corn, which harms or kills the plant. In the process of feeding, the beetles and their larvae can transmit viruses and bacteria that damage and/or kill various garden and crop plants. Because southern corn rootworms have destroyed a significant amount of the North America corn crop, Monsanto, DuPont, and Dow Chemical have all developed genetically engineered (GMO) corn. The roots of GMO corn synthesize a protein that is toxic to southern corn rootworms. Studies have shown that if in some way the protein was consumed, it would not harm people or livestock. GMO corn has enabled farmers to be far less reliant on insecticides. Unfortunately, in some areas, southern corn rootworms have developed resistance to the GMO corn.

SUMAC FLEA BEETLE

Latin name: *Blepharida rhois*

Family: Chrysomelidae

Identification: 0.2 to 0.3 inch long, convex oval, orange head and pronotum, elytra striped red and white

Distribution: Ontario to Florida, west to Texas

Comments: Both adult and larval sumac flea beetles feed on the leaves of bushes and trees in the sumac (Anacardiaceae) family. Sumac bushes and trees are protected by a poison; however, adult and larval sumac flea beetles are immune to the poison. Larval sumac flea beetles pile their frass that contains the sumac toxin on top of their bodies. The frass attaches and renders the beetle larvae poisonous. Many kinds of plants synthesize poisons for protection against insects. Some insect species that feed on these plants are immune to the poison and, like the sumac flea beetle, use the poisonous insecticide in various ways for their own protection. People also employ some of these insecticides to protect their gardens and commercial crops. Although the plant insecticides that are used by gardeners and farmers are synthesized commercially, some are identical to the insecticides that are made by plants and, therefore, are permitted to be used on organic farms.

THISTLE TORTOISE BEETLE

Latin name: *Cassida rubiginosa*

Family: Chrysomelidae

Identification: 0.2 to 0.6 inch long, green, looks like a miniature tortoise

Distribution: Nova Scotia to Virginia, west to Alberta and South Dakota

Comments: The top of the tortoise beetle's exoskeleton looks like the top a turtle shell, hence the name tortoise beetle. One of the main predators of leaf beetles are ants, but ants can't get a grip on the smooth, hard exoskeleton of tortoise beetles. This leaves the bottom (ventral) surface vulnerable. However, tortoise beetles secrete an oil that covers the bottom of their feet, enabling them to temporarily stick to the leaf that they are on. After attempting to pry off the beetle with its mandibles, the ant usually gives up and walks away.

The little green thistle tortoise beetle was brought to Europe from Asia to control thistles and was subsequently imported from Europe to the United States and Canada. The beetle feeds primarily on Canadian thistle (*Cirsium arvense*), a perennial weed that is spread by seeds and underground roots and tends to take over gardens and fields.

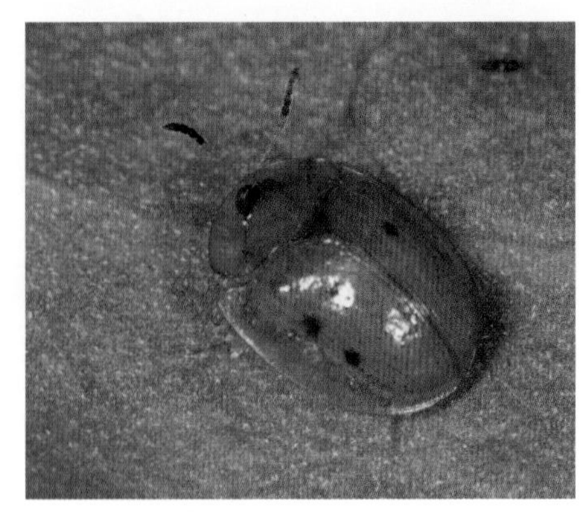

GOLDEN TORTOISE BEETLE, GOLDENBUG

Latin name: *Charidotella sexpunctata*

Family: Chrysomelidae

Identification: 0.2 to 0.3 inch long, dome-shaped, color can be golden or reddish with black spots, margins of elytra are transparent

Distribution: Throughout the United States and southern Ontario

Comments: Golden tortoise beetles are not always golden. In fact, they are one of the very few insects that can rapidly change color. They change color between gold and red as they age, when they mate, and when they are disturbed. The golden color fades when the beetle dies. They feed on morning glory and sweet potato vines, and other plants in the morning glory family. In the spring, females lay clusters of eggs on the undersides of leaves. When they hatch, the larvae place frass on their body for camouflage. The frass accumulates and also covers them when they pupate. Adults emerge in the late summer and overwinter.

GOLDENROD LEAF MINER

Latin name: *Microrphopala vittata*

Family: Chrysomelidae

Identification: 0.2 to 0.25 inch long, dark red with rows of punctures on elytra

Distribution: Widely distributed across southern Canada and the United States

Comments: Goldenrod leaf miners spend their entire lives on goldenrod. The beetles are called miners because their young larvae mine (feed on) the tissue between the top and bottom layers of the epidermis of goldenrod leaves. You might think that an insect larva is too big to fit inside a leaf, but the larvae of many different kinds of insects are miners when they are young. Inside the leaf, they are warmed by the sun and hidden from predators.

Goldenrod leaf miners are considered a keystone species—animals that have a significant effect on the environment even though there are relatively few of them. Goldenrod is a fast-growing tall plant that often takes over a field and shades out lower plants. From time to time, goldenrod leaf miners kill off most of the goldenrod in an area. This allows smaller diverse species to populate the field. Usually, it takes a number of years for the goldenrod to become reestablished.

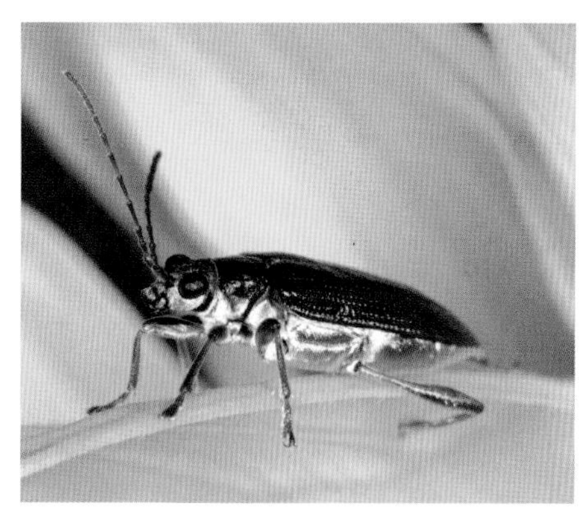

AQUATIC LEAF BEETLES

Latin name: *Donacia* sp.

Family: Chrysomelidae

Identification: 0.2 to 0.3 inch long, metallic brown or dark green, head and thorax narrower than abdomen, long antennae, thick femur

Distribution: Eastern North America, west to the Rocky Mountains

Comments: Adult leaf beetles of the genus *Donacia* feed on aquatic plants such as water lilies. They are alert, fast-flying little beetles but are not truly aquatic since they live and feed on the part of the plant that is above water. However, their larvae, which live and feed on the same plants, are aquatic, as they live underwater where they feed on the stems and roots. *Donacia* larvae breathe air from the stems and roots of the plant by thrusting the knifelike terminal segments of their abdomen into spongy tissue that contains air (aerenchyma). The terminal segments are connected to the larvae's trachea.

VIBURNUM LEAF BEETLE

Latin name: *Pyrrhalta viburni*

Family: Chrysomelidae

Identification: 0.2 to 0.3 inch long, shiny yellowish brown

Distribution: Southern Canada and northern United States

Comments: Female viburnum leaf beetles excavate rows of little holes in the twigs of viburnum plants and deposit several eggs in each one before covering the eggs with frass, chewed bark, and mucous. This cap hardens, keeping the eggs moist and helping protect them from predators. Larvae undergo three molts before falling to the ground, where they pupate and emerge as adults in about 10 days.

These Eurasian beetles were first discovered in North America in Ontario in 1947 and since then have spread to the northern part of North America from coast to coast. They do not appear to be extending their range to southern states, probably because they overwinter as eggs that need a prolonged period of cold weather to develop properly. Viburnum leaf beetles feed on many of the approximately 150 species of viburnum shrubs, some of which are cultivated as decorative bushes for yards and gardens. Both larvae and adults feed on the leaves of the plant. The leaves fall off, and the shrub is damaged and sometimes killed.

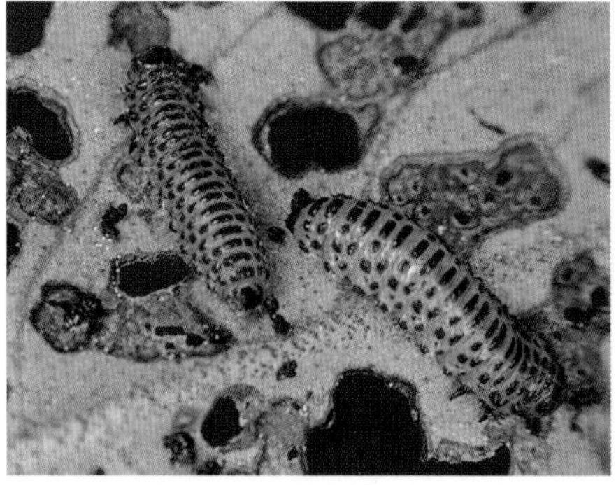

Viburnum leaf beetle larvae feeding on a viburnum leaf

WARTY LEAF BEETLE

Latin name: *Neochlamisus bebbianea*

Family: Chrysomelidae

Identification: 0.1 to 0.15 inch long, stout, shiny reddish brown, coarsely pitted with sharp bumps

Distribution: Maritime Provinces to Florida, west to Manitoba and Texas

Comments: Warty leaf beetles are stout and rough in appearance, which enables these little beetles to pass for caterpillar frass. Adult females lay their eggs on their host plant and cover them with frass. When the larvae hatch, they continue to cover themselves with their own frass in order to construct a case for protection. The case is open at one end so that the larva can extend its head and thorax as it walks and feeds on the plant. The case of these beetles looks like caterpillar frass, so both adults and their larvae are usually not attacked by predators. When they have matured enough to molt, the tiny larva seals the case with frass and pupates inside. The adult beetle then chews a hole in the case and walks out ready to fly and feed. Several species of case-bearing leaf beetle larvae form cases that look like caterpillar frass.

Warty leaf beetle larva in its case

STRIPED CUCUMBER BEETLE

Latin name: *Acalymma vittatum*

Family: Chrysomelidae

Identification: 0.2 to 0.3 inch long, black head, yellow thorax, yellow elytra with 3 wide longitudinal stripes

Distribution: Throughout North America

Comments: Striped cucumber beetles are a serious pest of the cucumber family, including cucumbers, squashes, pumpkins, zucchini, and melons. The beetles overwinter as adults. When they emerge in the spring, they feed on the cotyledons, stems, leaves, pollen, flowers, and fruits. Females lay their eggs next to plants. When they hatch, the larvae feed on roots and stems. Feeding by the adults and larvae can kill young plants and defoliate larger plants. However, the worst damage to plants is a disease called bacterial wilt, caused by a bacterium, *Erwinia tracheiphila,* which is transmitted by the beetles. Once the plant is infected, it wilts and dies, and there is no cure. Fruit is usually not harvested from plants infected with *Erwinia tracheiphila* because the quality of the fruit is so poor that it is unmarketable.

Tiger beetle, *Cicindela sexguttata*

TIGER BEETLES

Latin name: *Cicindela* sp.

Family: Carabidae

Identification: 0.4 to 0.6 inch long, huge eyes and mandibles, long legs

Distribution: Various species throughout North America, especially on beaches; some species threatened

Comments: Almost all species of tiger beetles have a similar lifestyle. Many species are colorful and iridescent. They live in sunny open areas, particularly beaches and dirt roads, where they run and fly rapidly. The predacious adults hunt on hot summer days, while the larvae live in small, short, narrow, cylindrical-shaped burrows that they dig in the sand or dirt. Larvae wait for an insect to come near enough that they can grab and devour it.

Tiger beetle, *Cicindela punctulata*

Here on Cape Cod, adults prowl the beaches on hot summer days, and they are not easy to catch. I bring my net to the beach, and when I see a tiger beetle, I approach cautiously. When I am about 8 or 10 feet away, the beetle flies up the beach and lands a couple dozen feet away facing me. When I approach again, the beetle flies off once more and lands further up the beach facing me. After repeating this several times, the beetle flies off into the distance.

BROAD-NECKED ROOT BORER

Latin name: *Prionus laticollis*

Family: Cerambycidae

Identification: 0.8 to 1.3 inches long, dark brown, eyes widely separated, females flightless

Distribution: Quebec to Florida, west to Minnesota and Texas

Comments: The larvae of this rather ugly longhorned beetle bore into roots, potentially causing damage to the trees and other plants they feed on. The nocturnal adults hide during the day by burrowing into the ground or hiding at the base of trees. Females lay their eggs in the soil or under leaf litter, usually in groups of twos, threes, or fours. After the eggs are laid, the females move their ovipositor up and down to fill the hole so it will not be seen by predators. When the larvae hatch, they dig into the ground, where they employ their strong mandibles to chew the roots on which they feed. They remain in the ground for 3 years before pupating and emerging from the soil as adults.

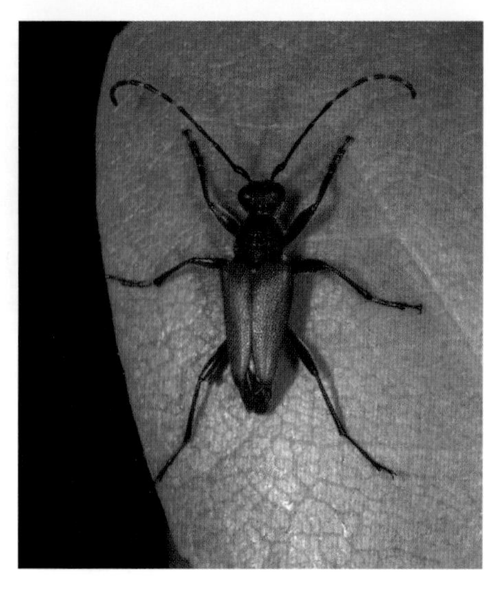

DISTENIID LONGHORNED BEETLE

Latin name: *Brachyleptura champlaini*

Family: Cerambycidae

Identification: 0.3 to 0.4 inch long, uniformly brown to reddish elytra, black head and thorax

Distribution: Nova Scotia to Florida, west to Ontario and Tennessee

Comments: Disteniid longhorned beetles and their relatives are called flower longhorned beetles because they are usually found feeding on the nectar and pollen of flowers. They are slender beetles and look a lot like wasps when they fly. Adult disteniid longhorned beetles are usually found in late spring and early summer on various flowers, especially New Jersey tea (*Ceanothus americanus*), spirea (*Spiraea*), aruncus (*Aruncus*), chestnut (*Castanea*), and beebalm (*Monarda*). Because they spend much of their time on flowers, they are considered minor pollinators. Larvae develop in decaying wood.

Longhorn beetle on a primrose, accompanied by smaller tumbling flower beetles

FLOWER LONGHORN BEETLE

Latin name: *Strangalia luteicornis*

Family: Cerambycidae

Identification: 0.4 to 0.6 inch long, elongate with pointed abdomen, yellow or yellowish brown, black spots on each side of head, 2 black bars on pronotum, long legs

Distribution: Ontario to Florida, west to Arkansas, Minnesota, Kansas, and Texas

Comments: Beetles in the family Cerambycidae are sometimes called false longhorn beetles. They are active during the day and usually found in the spring and summer on flowers, especially umbrella flowers such as buckweed, redbay, fleabane, holly, celery, carrot, and Queen Anne's lace. Larvae are wood borers and feed on the wood of various shrubs and hardwoods. Because they usually live in dead and dying trees, they help to recycle organic material.

ZEBRA LONGHORN BEETLE

Latin name: *Typocerus zebra*

Family: Cerambycidae

Identification: 0.4 to 0.6 inch long, elongate, black with distinct yellow markings, elytra narrows posteriorly

Distribution: New Jersey to Florida, west to Ohio and eastern Texas

Comments: Unlike most beetles that are oval-shaped and have hard elytra, *Typocerus zebra* and other flower longhorns are long, streamlined, cigar-shaped beetles. With their long shape and black and yellow coloring, flower beetles are pretty convincing wasp mimics. They feed on pollen and nectar and can usually be found on flat, open plants, particularly Queen Anne's lace, dogwood, or hydrangeas, and on roses. They sometimes are covered with pollen and are considered minor pollinators.

BRANCH PRUNER

Latin name: *Psyrassa unicolor*

Family: Cerambycidae

Identification: 0.3 to 0.5 inch long, slender, uniformly brownish yellow, females larger than males

Distribution: Quebec and Ontario to South Carolina, west to Texas

Comments: Adult branch pruners emerge from the trees where they pupated in the spring and early summer. Females mate and lay their eggs on small twigs. When the larvae hatch, they tunnel down the twig to a larger branch, bore into the branch, and start to excavate tunnels that encircle the branch. As they grow and molt, the larvae completely encircle the branches with concentric tunnels. When the larvae are ready, they move to just below the bark and pupate. Usually, the branch doesn't break off until winter, when it falls and the smooth cuts look almost like the branch has been pruned. The damage to the oak, hickory, and pecan trees that the beetle attacks is usually minor.

RED MILKWEED BEETLE

Latin name: *Tetraopes tetrophthalmus*

Family: Cerambycidae

Identification: 0.3 to 0.6 inch long, bright red with black markings

Distribution: Throughout North America, wherever there are milkweed plants

Comments: In early summer, female red milkweed beetles lay their eggs at the base of milkweed stems. When the larvae hatch, they dig into the ground and feed on the roots of the milkweed plant until the fall. These larvae overwinter in the roots until the spring, when they pupate and emerge as adults that feed on the leaves, buds, and flowers of the plant. Because milkweed leaves contain sticky latex, red milkweed beetles constantly rub their mandibles against the plant to remove the latex before it hardens. Red milkweed beetles do not use pheromones to find mates—they just look for milkweed plants, if they are not already on one. If there are a lot of females on the plant, a male will usually choose to mate with the larger ones. However, if there are a lot of males and few females, the males will become competitive, with the females usually choosing to mate with the larger ones.

RED OAK BORER

Latin name: *Enaphalodes rufulus*

Family: Cerambycidae

Identification: 0.6 to 1 inch long, reddish brown with patchy golden setae

Distribution: Throughout North America, wherever oak trees are found

Comments: Although this species it is called the red oak borer, it also attacks other species of oaks, especially black and scarlet oaks. Damage is caused by the tunnels that the larvae chew in the wood and by other insects that take up residence in the tunnels. Oak lumber from infested trees is worth far less because of the splinters, discoloration, and exit holes caused by the red oak borer larvae. The beetle has a 2-year life cycle. Interestingly, red oak borers are synchronized such that adults are only found in odd-numbered years. Adults mate on oak trees, and the female lays about a hundred eggs in cracks in the bark or under lichen patches. As soon as the larvae hatch, they bore through the bark and spend their first year in the sapwood. Second-year larvae tunnel deeper into the heartwood and pupate in their tunnels.

LOCUST BORER

Latin name: *Megacyllene robiniae*

Family: Cerambycidae

Identification: 0.4 to 1 inch long, wasplike with yellow and black bands

Distribution: Throughout North America, wherever black locust trees are found

Comments: Locust borers are usually found on goldenrod, where they feed on nectar. In late summer, females can also be seen running up and down young black locusts trees (*Robinia pseudoacacia*), particularly those that have been stressed by drought or poor soil, searching for cracks in the bark to lay their eggs. The larvae hatch in the spring and bore into the trees, where they mature and pupate. The tunnels in the black locust trees that are made by the larvae damage the trees and are often inhabited by the fungus *Phellinus robiniae,* which also damages the trees. In recent years, homeowners have become fond of black locust trees, which are also planted for the restoration of strip mines and to prevent erosion. As a consequence, locus borers have become more common.

FALSE BLISTER BEETLES

Latin name: *Asclera* sp.

Family: Oedemeridae

Identification: 0.2 to 0.4 inch long, black with orange pronotum, long segmented antennae

Distribution: Maritime Provinces and Ontario, south to Florida, west to Wisconsin, Colorado, and Texas (more common along coasts)

Comments: False blister beetles, Oedemeridae, contain cantharidin, the same blister-causing agent that is found in blister beetles, family Meloidae. False blister beetles feed on pollen. Their larvae develop in moist decaying stumps, logs, and tree roots. Larvae of an imported species, *Nacerdes melanura,* damage wharf pilings. Adult false blister beetles can also cause problems because they are attracted to lights. One species, *Oxycopis mcdonaldi,* which is found in large numbers in the Florida Keys, is attracted to lights around swimming pools, tennis courts, and open-air restaurants. If they are pinched or squashed against the skin, they release cantharidin, causing blistering.

Male American oil beetle

AMERICAN OIL BEETLE

Latin name: *Meloe impressus*

Family: Meloidae

Identification: 0.3 to 0.6 inch long, shiny metallic blue, green, or black; females larger than males

Distribution: Throughout North America

Comments: Male American oil beetles sport strange-looking antennae that they employ to hold the larger females when they mate. Holding on to your mate is very important in the insect world because females may walk or fly away—in this case, walk away, because American oil beetles cannot fly. Although these beetles are soft-bodied and flightless, they are protected by poison. A member of the blister beetle family, Meloidae, the American oil beetle, as well as some related beetles, contains the poison cantharidin, which causes blisters. Cantharidin obtained from the European blister beetle (*Lytta vesicatoria*), commonly called the Spanish fly, has been used historically as an aphrodisiac. It must have sickened a lot of people, as cantharidin is extremely toxic. One two-hundredth of an ounce is the lethal dose for humans. Cantharidin is such a powerful toxin that it can sicken and even kill horses that consume plants with large numbers of blister beetles on them.

BOSTRICHID BEETLE, POWDER-POST BEETLE

Latin name: *Lichenophanes bicornis*

Family: Bostrichidae

Identification: 0.3 to 0.4 inch long, elongate cylindrical, reddish brown, head smaller than pronotum

Distribution: Throughout eastern North America and southern Canada, east of Manitoba

Comments: Bostrichid beetles are also called powder-post beetles because their larvae leave a fine wood dust mixed with frass in the chambers that they make in the tree branches where they live. Unlike most beetles, bostrichid beetles are cylindrical in shape. One of the attributes that has facilitated the success of insects is that they have evolved into different shapes. The cylindrical shape of these beetles is perfect for living in the tunnels that they excavate in wood. *Lichenophanes bicornis* adults and larvae prefer various hardwoods. The larvae do most of the damage, and because they live in hardwood, hardwood furniture often harbors beetle larvae unless the wood has been heated or treated with chemicals that kill insects. This is probably how bostrichid beetles and other beetles and beetle larvae that live in wood have traveled from continent to continent.

EMERALD ASH BORER

Latin name: *Agrilus planipennis*

Family: Buprestidae

Identification: 0.3 to 0.4 inch long, slender, bright iridescent green

Distribution: Northeastern and upper Midwest of the United States, southern Ontario and Quebec

Comments: Emerald ash borers are members of the metallic wood borer family Buprestidae. Buprestidae are also called jewel beetles because of their bright iridescent colors. Unlike most beetles, metallic wood borers are fast and agile fliers. This iridescent green Asian species was first discovered in Ontario in 2002 and has spread throughout North America. In their native forests, Asian emerald ash borers don't cause much damage because they are kept in check by natural predators. However, in North America, emerald ash borers are devastating ash trees.

METALLIC WOOD-BORING BEETLES, JEWEL BEETLES

Latin name: *Taphrocerus* sp.

Family: Buprestidae

Identification: 0.1 to 0.15 inch long; shiny black with small white markings; setae on elytra, although they often rub off

Distribution: Eastern North America

Comments: Like other metallic wood-boring beetles, the genus *Taphrocerus* fly fast and maneuver rapidly. These little beetles feed on the margins of sage leaves in the spring and summer, and they can usually be found on sages that grow along the edges of wetlands, roadsides, and ditches. Females lay their eggs in the leaves of the sage plants. When the larvae hatch, they feed (mine) on the tissue between the upper and lower epidermis of the leaves, where they cannot be seen by predators. There are one or two generations each year.

RED-NECKED CANE BORER

Latin name: *Agrilus ruficollis*

Family: Buprestidae

Identification: 0.15 to 0.3 inch long, elongate cylindrical, shiny bluish or black, red pronotum

Distribution: Throughout the United States and southern Canada

Comments: These jewel beetles tend to be active on hot summer days. Female red-necked cane borers lay eggs on the bark of raspberry, blackberry, and dewberry bushes. When the eggs hatch, the larvae immediately bore spiral chambers under the bark, which in turn produce galls. Because the larvae live in galls, they are not exposed to the outside environment, and pesticide spraying cannot reach them. In the fall, mature larvae form cells in the pith, where they overwinter before pupating in the spring and emerging as adults through D-shaped holes. Adults are usually found on flowers. Red-necked cane borers have one generation a year.

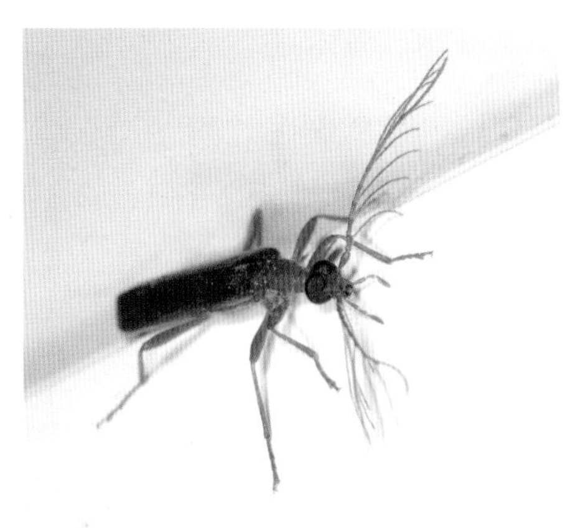

FIRE-COLORED BEETLE

Latin name: *Dendroides concolor*

Family: Pyrochroidae

Identification: 0.4 to 0.6 inch long, usually orange, black eyes, large branching antennae

Distribution: Newfoundland and Nova Scotia, south to North Carolina, west to Tennessee, Illinois, and Kansas

Comments: These nocturnal beetles usually hide beneath the leaves of trees and shrubs during the day. Their orange color warns predatory insects to stay away because they are poisonous. Their poison, cantharidin, is also the poison of blister beetles and false blister beetles. Male fire-colored beetles, as well as their blister beetle relatives, transfer cantharidin to females along with sperm when they mate. As the eggs pass down the female's oviducts, they become coated with cantharidin, which protects them from predators. Although cantharidin is extremely toxic and causes skin blisters, when it is properly dosed, it has been used to treat minor skin conditions.

FIREFLY, LIGHTNING BUG

Latin name: *Photuris pennsylvanica*

Family: Lampyridae

Identification: 0.3 to 0.4 inch long, slender, black pronotum flanked by reddish-brown and yellow spots

Distribution: Maritime Provinces and Quebec to Georgia, west to Ontario and Texas

Comments: Male fireflies use light to attract mates. *Photuris* females produce light flashes in response to males of their species to indicate where they can be found. Each species has a unique pattern of flashes. There are a few other luminescent insects, but they light up continuously. Perhaps the production of light is not common in insects because it could attract predators. However, fireflies are protected by poison, so their light flashes more likely serve to warn predators to stay away. The toxins, called lucibufagins, are chemically related to the cardiotoxins that are produced by toads and certain poisonous plants. Females of some species of the genus *Photuris* can mimic the flashes of certain other fireflies species. When a male *Photuris* follows the signal of a female of his own species, they mate. However, when a male of a species that the *Photuris* female is mimicking follows the signal to her, she kills and eats him.

Winter fireflies mating in early spring

WINTER FIREFLY

Latin name: *Ellychnia corrusca*

Family: Lampyridae

Identification: 0.4 to 0.6 inch long, oval, dark black, yellow and orange stripes on pronotum

Distribution: Maritime Provinces to northern Florida, west to Manitoba and North Dakota

Comments: These fireflies overwinter as adults and in late winter and early spring can be found mating on trees. Although they are bioluminescent as larvae and young adults, they lose their ability to produce light as they grow older. Therefore, males probably find mates by following the concentration gradient of sex pheromones that are released by females. Their mating ritual involves the male touching the female with his antennae and mouthparts. The couple then moves back-to-back and the male transfers a sperm packet to his mate. Like other fireflies, winter fireflies are protected by poisons that they secrete from their leg joints.

JAPANESE BEETLE

Latin name: *Popillia japonica*

Family: Scarabaeidae

Identification: 0.4 inch long; coppery green head, pronotum, and legs; reddish-brown elytra

Distribution: Throughout North America, except West Coast

Comments: These shiny green-and-orange beetles are native to Japan, where they don't cause much of a problem because they are controlled by natural predators. However, since their introduction into North America in the early twentieth century, Japanese beetles have caused immense damage to lawns, golf courses, fields, gardens, and crops, where adult beetles feed on the leaves of nearly 300 different plants, including fruit and shade trees. Adult females lay their eggs in holes that they dig in grassy places like lawns, parks, and golf courses. Larvae feed on the roots of the grass, which can leave lawns unsightly. Japanese beetle larvae are a favorite food of skunks that dig them up, rendering the lawns and golf courses even more unsightly. Japanese beetles are widespread east of the Rocky Mountains but not found west of the Rockies, probably because the mountains form a natural barrier. Farmers and gardeners in the western states are, of course, eager to keep them out.

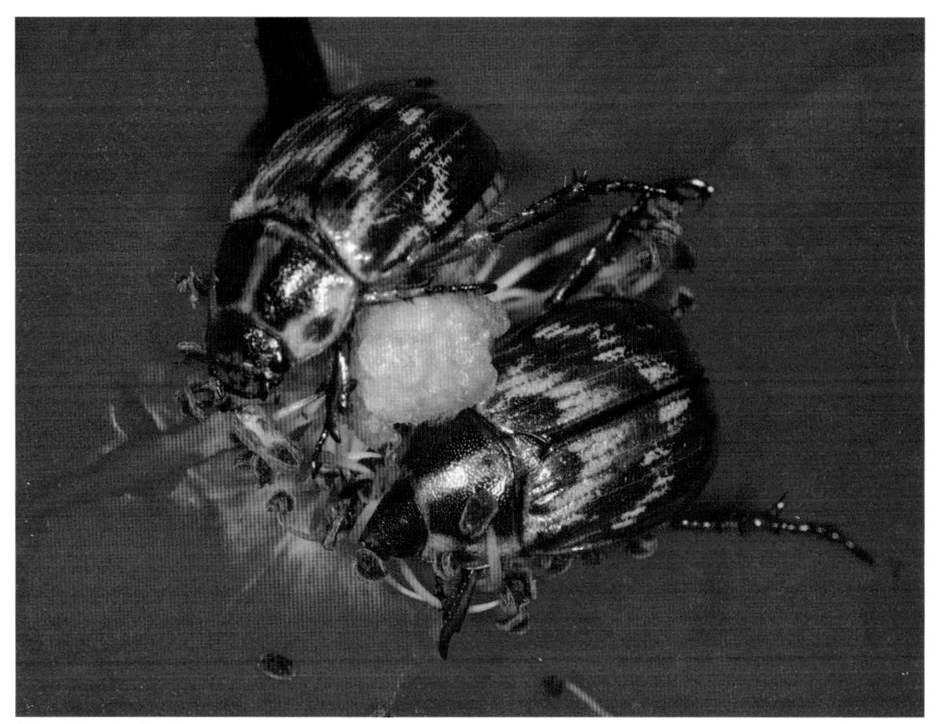

Oriental beetles feeding on a rose

ORIENTAL BEETLE

Latin name: *Anomala orientalis*

Family: Scarabaeidae

Identification: 0.3 to 0.4 inch long; mottled metallic brown or black head, thorax, and elytra

Distribution: Maine, south to South Carolina, west to Wisconsin

Comments: This pest was imported sometime before 1920, which is when it was first detected in Connecticut. It is sometimes confused with the Japanese beetle because they are both about the same size scarab beetles; however, the Japanese beetle has reddish brown elytra and a coppery green head and thorax, while the Oriental beetle is mottled brown. Although the Japanese beetle is a more serious pest, they both have the same lifestyle. Adults feed on flowers and leaves, and their larvae feed on the roots of grasses and can cause major damage to lawns, golf courses, and parks. There are pheromone traps for Oriental beetles available that are safe to use because they do not contain insecticides. The traps are also inexpensive. If you buy one, in essence, you will get two for the price of one, because they also trap Japanese beetles.

GRAPEVINE BEETLE

Latin name: *Pelidnota punctata*

Family: Scarabaeidae

Identification: 1 inch long, yellow or tan, usually with 3 black spots on elytra and 2 on pronotum

Distribution: Ontario to Florida, west to South Dakota and Texas

Comments: There are many variations in the color pattern of these beetles. When they were first described in 1915, the color variations were believed to be different species. Although the insects can identify members of their own species regardless of the color variations among individuals, it can be very confusing for insect enthusiasts. Adult grapevine beetles feed on grape leaves and grapes, but they are generally not numerous enough to cause much damage. Their larvae usually live in decaying wood. If you don't live where grapes are grown, you may still see these beetles because in addition to commercially grown grapes, grapevine beetles feed on the leaves of the many species of wild grapes. The beetles fly rapidly on warm nights and are attracted to lights. Grapevine beetles do not bite and are harmless.

Female rhinoceros beetle. Females do not have horns.

RHINOCEROS BEETLE, HERCULES BEETLE

Latin name: *Xyloryctes jamaicensis*

Family: Scarabaeidae

Identification: 0.8 to 1.3 inches long; dark reddish brown with long, dense setae underneath; males have a horn on the pronotum

Distribution: Ontario to Georgia, west to Arizona

Comments: Rhinoceros beetles have powerful legs that they use for digging. Although they can fly, their flight is awkward, so their best defense is their large size and solid build. They are called rhinoceros beetles because males of some species have a hornlike structure on their head or pronotum, which they employ to joust with other males for mating rights. To test the strength of these beetles, scientists have placed weights equivalent to 800 times the weight of the beetle on their backs. The beetles could still walk while carrying this tremendous weight. In Asia, rhinoceros beetles are often kept as pets, as they are large and easy to maintain. If two males are put in a small cage with a female, they will joust for mating rights.

DUNG BEETLES

Latin name: *Onthophagus* sp.

Family: Scarabaeidae

Identification: 0.2 to 0.4 inch long, usually dark brown, chubby, males of some species have a horn on their thorax

Distribution: Throughout North America

Comments: Both larva and adult dung beetles feed on dung. As they are especially valuable to the cattle industry, different species have been imported from Europe to North America and Australia where cattle are farmed. The tunneling of dung beetles recycles the soil and increases the ability of soil to absorb water. Also, cattle will not graze where there is dung, so by removing the dung patties, these beetles render more of the fields available for grazing. Another advantage of removing dung is that dung is the breeding ground for pests such as the horn fly (*Haematobia irritans*), which feeds on the blood of cattle, weakening them. Another pest that feeds on dung, the face fly (*Musca autumnalis*), transmits diseases of the eye to cattle. Thus, although you may find the lifestyle of dung beetles a bit disgusting, they can make cattle farms more productive.

Male scooped scarab beetle

SCOOPED SCARAB BEETLE

Latin name: *Onthophagus hecate*

Family: Scarabaeidae

Identification: 0.2 to 0.3 inch long, uniformly dull black, males have triangular upturned pronotum

Distribution: Eastern United States and southern Canada, west to eastern Texas and Colorado

Comments: Males have an extension of their pronotum called a clypeus, which is why they are called scooped scarab beetles. When structures like these are found on male but not female beetles, their function usually has to do with jousting other males for mating rights. Scarabaeidae is a large family of beetles; scooped scarab beetles belong to the subfamily Scarabaeinae. Both adults and young scooped scarab beetles usually feed on cow dung, although they also feed on the dung of other mammals and sometimes are found in rotting fungi, fruit, and carrion.

GROUND BEETLES

Latin name: *Calleida* sp.

Family: Carabidae

Identification: 0.4 to 0.5 inch long, thin and shiny with reddish thorax and green abdomen

Distribution: Massachusetts to Florida, west to Missouri

Comments: Ground beetles cannot fly. When threatened, they run for cover and can release a noxious cystic secretion from their anal glands. Darwin wrote an interesting little tale about an encounter he had with ground beetles in a letter to a friend that reads, "I must tell you what happened to me on the banks of the Cam in my early entomological days; under a piece of bark I found two carabi (I forget which) & caught one in each hand, when lo & behold I saw a sacred *Panagæus crux major* [a rare European ground beetle]; I could not bear to give up either of my Carabi, & to lose *Panagæus* was out of the question, so that in despair I gently seized one of the carabi between my teeth, when to my unspeakable disgust & pain the little inconsiderate beast squirted his acid down my throat & I lost both Carabi & *Panagæus*!"

MARGINED CARRION BEETLE

Latin name: *Oiceoptoma noveboracense*

Family: Silphidae

Identification: 0.5 to 0.6 inch long, black with orange-red margin on pronotum

Distribution: Maritime Provinces, south to Georgia, west to Mississippi, Oklahoma, and Montana

Comments: Margined carrion beetles are named for the orange-red margin on their pronotum. Like other carrion beetles, margined carrion beetles feed on dead animals, as well as the larvae of flies that have laid their eggs on carrion. These diurnal beetles mate in the spring. Males are often seen grasping the antennae of females with their mandibles and mounting them for some time, although they may not be mating. After riding the females, the males mate with them, but they do not leave their mates. Instead, they continue to hold on to their mates until they lay their eggs, presumably to ensure their paternity.

TOMENTOSE BURYING BEETLE

Latin name: *Nicrophorus tomentosus*

Family: Silphidae

Identification: 0.4 to 0.8 inch long, pronotum with dense setae, distinctive orange markings on black elytra

Distribution: Throughout southeastern Canada and North America to New Mexico

Comments: In the spring, pairs of tomentose burying beetles use their keen sense of smell to locate a dead animal—generally a mouse. They hide the dead mouse, usually by moving it to a shallow pit and covering it with leaf litter, and the female lays her eggs on or near the mouse. Flies also lay their eggs on dead mice so that when the maggots hatch, they too will consume the dead animal. The beetles eat the maggots, but if too many flies lay their eggs on the dead mouse, the abundance of maggots will consume the mouse faster than the beetles can eat them. However, the tiny symbiotic mites that live on tomentose burying beetles will come to their aid. The mites leave the beetles, feed on the flies' eggs and maggots, then jump back onto the beetles. In this way, they help reduce the number of maggots that compete for food with the beetles.

Symbiotic mites on a tomentose burying beetle

AMERICAN CARRION BEETLE

Latin name: *Necrophilia americana*

Family: Silphidae

Identification: 0.8 to 1.3 inches long, black with yellow pronotum surrounding black area

Distribution: Nova Scotia, south to Florida, west to Manitoba and eastern Texas

Comments: The dark color of these beetles probably helps them to blend in with the soil. American carrion beetles are active in spring and summer, mostly in damp woodlands. When they fly, they look like bumblebees. This ruse probably helps protect them from birds. These soft-bodied flat beetles and their larvae feed on dead animals and the larvae of flies that live on dead animals, although they also feed on fungi. Usually, these beetles bury the dead animal to prevent other insects from finding it. If the animal is large, a number of beetles will cooperate in burying the carrion. To catch these and other nocturnal ground beetles, I place a small amount of wet wheat germ in a jar and bury it in the ground such that the top of the jar is level with the ground. In the morning I look for beetles and other insects that were attracted by the wheat germ and fell into my trap.

TROGOSSITID BEETLE

Latin name: *Peltis septentrionalis*

Family: Trogossitidae

Identification: 0.4 to 0.5 inch long, flattened oval, shiny brown

Distribution: Across Canada and northern United States

Comments: This species of trogossitid beetle lives under the bark of dead trees, where they seek out the small beetles, beetle larvae, and eggs on which they feed. Insects have evolved into many different shapes so that they can live in so many different places; flattened insects like the trogossitid beetle can fit between the bark and wood of fallen trees. Numerous beetles and beetle larvae make their homes under the bark of trees, where they often make tunnels and chambers. The shape and arrangement of the tunnels are often characteristic for each family or genus of beetle. When trees die or are diseased, the bark often separates from the sapwood. Removing the bark from dead trees reveals the interesting tunnels and chambers that have been carved out by various species of beetles and their larvae.

TUMBLING FLOWER BEETLES, PINTAIL BEETLES

Latin name: *Mordella* sp.

Family: Mordellidae

Identification: 0.15 to 0.2 inch long, humpbacked wedge shape with a pointed abdomen

Distribution: Throughout the United States and southern Canada

Comments: Tumbling flower beetles live on flowers, especially those in the parsley family. Their larvae live in rotting wood or plant stems. When they are threatened, these tiny beetles tumble off the flower and play dead when they hit the ground or continue to tumble until they are in a good position to take off. The tumbling is actually a series of rapid jumps. If tumbling flower beetles are placed on a piece of paper, instead of flying or running away, they will tumble around.

WEDGE-SHAPED BEETLES, RIPIPHORID BEETLES

Latin name: *Ripiphorus* sp.

Family: Ripiphoridae

Identification: 0.1 to 0.3 inch long, greatly reduced elytra, looks like a fly

Distribution: Uncommon across North America

Comments: Wedge-shaped beetles are some of the most unorthodox-looking beetles in North America. The elytra of these beetles are reduced to small structures at the base of hindwings, making them look more like flies than beetles. Adult females lay their eggs on flowers. The eggs hatch almost immediately into small larvae that have legs, and the larvae attach to solitary bees that visit the flowers. Once inside the bees' nest, wedged-shaped beetle larvae consume the bee larvae. After consuming the bee larvae, they pupate in the nest and emerge as adult beetles. Adults only live a few days.

BEACH BEETLE

Latin name: *Phaleria testacea*

Family: Tenebrionidae

Identification: 0.2 to 0.3 inch long, light brown with darker brown spots on broadly notched abdomen

Distribution: Beaches along eastern coast of North America and Gulf Coast

Comments: Most species of North American insects inhabit a large range of the continent and many can be found in both fields and woodlands. However, some can only be found in one environment. For example, some species live on only one kind of tree and never leave the tree. The beach beetle, *Phaleria testacea*, in the family Tenebrionidae, is a kind of darkling beetle that lives under seaweed and sometimes driftwood or other objects on beaches. You won't find them anywhere else. As is the case of most darkling beetles, the beach beetle feeds on plant material; in this case, seaweed.

WHIRLIGIG BEETLES

Latin name: *Gyrinus* sp.

Family: Gyrinidae

Identification: 0.2 to 0.3 inch long, brown or black, elytra with 11 rows of grooves

Distribution: Throughout North America

Comments: Groups of hundreds of oval-shaped whirligig beetles zip around on the surface of ponds, lakes, and streams. Members of the group move from the outside of the raft to the inside and back. It is safer in the middle of the raft, but the water outside the group contains more tiny insects on which the beetles feed. These little aquatic beetles have two eyes above the surface of the water and two below. You might think that by living on the surface of the water these beetles would be an easy meal for hungry fish, but the whirligig beetle synthesizes a defense poison. When a fish attempts to eat a whirligig beetle, it detects the poison. In response, the fish retains the beetle in its mouth while taking water in, then expels the water through its gills to flush out the poison. However, whirligig beetles secrete their poison gradually for several minutes. When the fish still tastes the poison after a minute or so, it spits out the beetle.

Whirligig beetles have four eyes: two to see above the surface of the water and two below.

CHECKERED BEETLES

Latin name: *Enoclerus* sp.

Family: Cleridae

Identification: 0.3 to 0.4 inch long, elongated, pubescent, brown with prominent black and white stripes

Distribution: Throughout North America

Comments: Checkered beetles are hairy, colorful, cylindrical-shaped little beetles with bulging eyes. Checkered beetles of the genus *Enoclerus* are often found on the base of trees or under bark, where they feed on bark beetles and other borers. Bark beetles are a major pest that destroys millions of trees. If you see one of these little checkered beetles, you should be thankful. By consuming bark beetles along with their larvae and eggs, checkered beetles have saved millions of trees. Although most *Enoclerus* species feed on bark beetles, other kinds of checkered beetles feed on the caterpillars and larvae of other insects. Some species of *Enoclerus* visit flowers, where they lay their eggs. When the larvae hatch, they grab on to bees that visit the flowers and hitch a ride back to the bees' nest, where they feed on the bees' larvae. Other *Enoclerus* species mimic velvet ants.

CLICK BEETLE, JACKKNIFE BEETLE, SKIPJACK

Latin name: *Megapenthes limbalis*

Family: Elateridae

Identification: 0.3 to 1 inch long, slender

Distribution: Throughout North America

Comments: There are almost 1,000 species of click beetles in North America. They are named for their ability to flip into the air in order to right themselves when they fall on their back. They do this by snapping a spinelike process on their thorax into a groove. This makes an audible click sound and creates enough force to flip them into the air as far as several times the length of their body. In the event they happen to fall on their back again, the beetles repeat the process until they land on their feet. It is believed that the clicking sound and rapid leap into the air may also frighten would-be predators. The long, thin larvae of click beetles are called wireworms. Those of most species live in the ground and feed on roots.

GOLDENROD SOLDIER BEETLE, PENNSYLVANIA LEATHERWING

Latin name: *Chauliognathus pennsylvanicus*

Family: Cantharidae

Identification: 0.4 to 0.5 inch long, orange with 2 black areas on rear of elytra, black head and legs, black markings on thorax, elytra do not completely cover abdomen

Distribution: Eastern United States, east to Texas; southeastern Canada

Comments: Adult goldenrod soldier beetles feed primarily on pollen and nectar, especially goldenrod, but also sometimes devour aphids and other small soft-bodied insects. The larvae live in leaf litter, under the ground, or in decaying wood, where they feed on insect larvae and eggs. Both adults and larvae have glands at the rear of their abdomen that secrete defensive chemicals. Like other soldier beetles, they fly well, so well that they look a lot like wasps or bees when they are flying. Because they usually visit numerous flowers, these and other soldier beetles are minor pollinators.

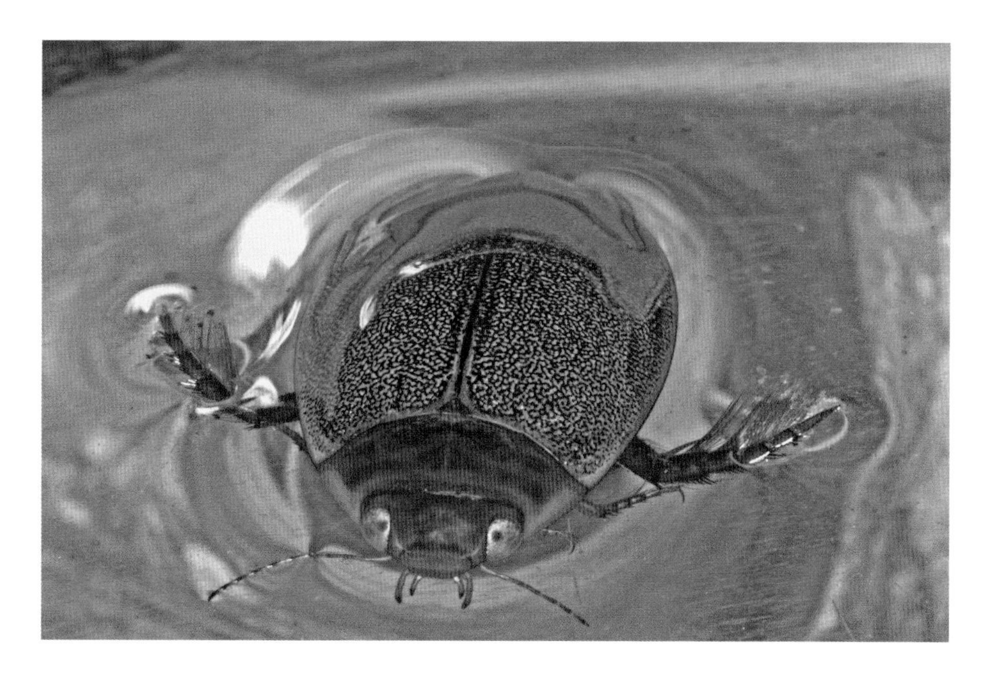

PREDACEOUS WATER BEETLES, PREDACEOUS DIVING BEETLES, TRUE WATER BUGS

Latin name: *Rhanthus* sp.

Family: Dytiscidae

Identification: 0.4 to 1 inch long, brown or black, although some are reddish brown

Distribution: Throughout North America, primarily in the east

Comments: These aquatic beetles swim by moving their large flattened and fringed hind legs in unison like oars. When they are resting, they usually move the back of their abdomen above the surface of the water and breathe through their spiracles. However, when preparing to dive, they turn over and collect an air bubble from the surface and move it under their elytra. Predaceous water beetles fly well and sometimes fly to swimming pools and puddles. They usually hold on to aquatic vegetation, waiting for an insect or small pollywog or fish to pass by. They then lunge from the resting place and swim rapidly in pursuit of their prey. The mandibles of these beetles are small, but they are sharp, and the beetle can deliver toxins that paralyze their prey. The front legs of male predaceous diving beetles are equipped with suction pads, which they use to attach their front feet to females during mating.

Suction pads on the front feet of the male predaceous diving beetle are used to hold the female while mating.

ROVE BEETLES

Latin name: *Ocypus* sp.

Family: Staphylinidae

Identification: 0.4 to 0.6 inch long, black, short elytra, flexible abdomen

Distribution: Throughout North America

Comments: Rove beetles are the largest family of beetles in North America with over 4,000 species, most of which can only be identified by experts. They are fast-moving, usually shiny, elongated beetles that are often mistaken for earwigs. Both adults and larvae are predacious. Rove beetles are found in many habitats including dung, carrion, and ant colonies. Some species live on the shores of oceans, lakes, and ponds and on the perimeter of ant colonies, but most live in leaf litter or under stones. The elytra of rove beetles are so short that they only cover the most anterior portion of the abdomen. The beetles can lift their elytra, unfold their hindwings, and fly quite well. However, they usually prefer to run away when threatened or turn around, lift their abdomen over their body or to the side, and spray defensive chemicals at an attacker.

22 NEUROPTERA

Neuropterans are delicate soft-bodied insects that have four long wings with many veins, biting mouthparts, and short legs. Adults of many species resemble dragonflies. However, unlike dragonflies, neuropterans are weak fliers and undergo complete metamorphosis. Neuropteran larvae have large biting mouthparts (jaws) with a groove through which liquid can pass. Their jaws are not only for biting but for injecting and sucking. When these predatory larvae bite, they inject toxins that paralyze their prey and enzymes that digest the victim's tissues. The liquefied tissues are then sucked up through the groove in the jaws. Larvae of many neuropterans have an anchoring device with little hooks that they can extrude from their anus. The device helps to secure them when they are being pulled by large prey.

GREEN LACEWINGS

Latin name: *Chrysopa* sp.

Family: Chrysopidae

Identification: 0.4 to 0.5 inch long, light green, delicate, clear wings with many veins, large gold-colored eyes

Distribution: Throughout the United States and southern Canada

Comments: Most neuropterans are uncommon, but the green lacewing is an exception. Various species of these beautiful delicate insects are common throughout North America. Adults are usually omnivores, feeding on pollen nectar and small insects. On the other hand, green lacewing larvae are strictly predacious and feed on small insects, including aphids, mealybugs, and thrips, and insect eggs. They have very large mandibles that have a groove on the inside through which the larva can secrete venom that immobilizes its prey. Green lacewing larvae often pile bits of grass and leaves on top of themselves so that predators will not see them, especially ants that often guard the aphids on which lacewing larvae feed. All in all, green lacewing larvae have earned the name "aphid lion" and are one of the gardener's and farmer's best friends. In fact, they are raised and sold to help rid gardens and farms of aphids. They are favored by many gardeners and farmers over lady beetles because, unlike lady beetles, the adults are weak fliers and tend to stay put. Gardeners plant sunflowers and certain other plants to attract green lacewings.

Green lacewing larva

To attract a mate, adult green lacewings make sounds. Although the sound is not loud enough for humans to hear, green lacewings have an auditory organ at the base of their forewings that enables them to hear the call of a nearby lacewing of the same species. The sound made by different species is slightly different. Lacewings are active both day and night, and they can hear the ultrasonic calls of bats. If they hear a bat when they are flying, they fold their wings, which makes their echolocation signal smaller, and fall to the ground where bats cannot detect them.

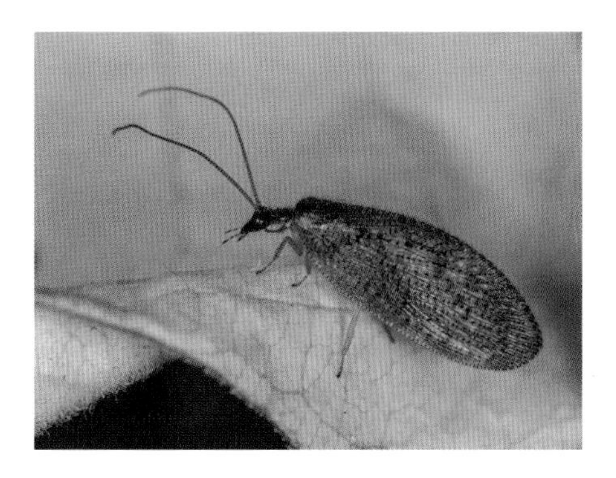

BROWN LACEWINGS

Latin name: *Hemerobius* sp.

Family: Hemerobiidae

Identification: 0.2 to 0.3 inch long, dark to light brown, wings with short hairs, resemble green lacewings but smaller and thinner

Distribution: Throughout North America and southern Canada

Comments: Brown lacewings look like green lacewings, but they are brown, smaller, and less numerous. They usually overwinter as pupae, but they can sometimes overwinter as adults. They can often be seen in the spring, not only because they overwinter as adults, but also because they are tolerant of cold weather. Brown lacewings are predacious as adults and larvae and are therefore sometimes called "aphid wolves." Because they feed on soft-bodied insects and mites, especially aphids, they are sold to gardeners to control aphids and mealybugs. Females lay their eggs directly on the stems and leaves of plants unlike green lacewings, which create stiff silk stalks on which to lay each egg. Brown lacewings are attracted to lights. They can also be found by sweeping a net through grasses.

BEADED LACEWINGS

Latin name: *Lomamyia* sp.

Family: Berothidae

Identification: 0.2 to 0.3 inch long, brown hairy legs and wings, outer margin of forewings indented

Distribution: Throughout North America

Comments: Female beaded lacewings lay their eggs adjacent to dry-wood termite nests. When the larvae hatch, they crawl into the termite nest and feed on the termites. They can discharge a poisonous gas from their anus that immobilizes the termites, which are then gobbled up by the hungry beaded lacewing larvae. To protect their nest, some termites in the colony are "soldiers" equipped with powerful mandibles, but the soldiers ignore beaded lacewings. Because most termites are blind and it is dark in the colony, termites identify one another and other insects by the surface molecules on their exoskeleton. However, many insects that live in termite and ant colonies are protected by their hosts or, like beaded lacewings, feed on their hosts. Most of these insects have evolved surface molecules that are similar, or identical, to the surface molecules of their termite or ant hosts so as not to be identified as aliens.

SPOTTED ANT LION, DOODLEBUG, ANTLION LACEWING

Latin name: *Dendroleon obsoletus*

Family: Myrmeleontidae

Identification: 1.5- to 2-inch wingspan, mottled-pattern wings

Distribution: Northeastern United States

Comments: At first glance, adult ant lions look like dragonflies, but they have relatively long antennae with a club at the tip, whereas dragonflies have very short, thin antennae. Unlike dragonflies, ant lions are feeble and awkward fliers. Also, dragonflies are diurnal, and ant lions are nocturnal. Ant lions are named for their larvae, which are very formidable predators. Some ant lions dig funnel-shaped traps in sand or soft soil, whereas the larvae of most species hide under plants in shallow burrows in the ground, where they capture insects that walk by. The ant lions hide at the bottom of the pit and wait for an ant or other insects to fall into their trap. The larvae are covered with numerous hairs that anchor them firmly to the soil at the bottom of the pit, with only their prominent mandibles protruding. When an insect falls into the pit, the ant lion skewers it with its mandibles and paralyzes it with toxins that are secreted into a groove in its mandibles, before consuming the helpless insect.

DOBSONFLY

Latin name: *Corydalus cornutus*

Family: Corydalidae

Identification: 2 to 2.5 inches long, 4- to 5.5-inch wingspan, forewings with dark veins, males have very long mandibles crossed at tips, females have short mandibles

Distribution: Depending on species, throughout the United States and southern Canada

Comments: Although males of these large neuropterans have long mandibles, they cannot bite. Females have far shorter mandibles but can give a painful bite. The mandibles of males are used to flip other males into the air when they compete for mating rights. After a male successfully competes, he approaches the female and touches her with his antennae. If she accepts him, while they mate, the male attaches a "nuptial gift" of a large nutrient-rich spermatophore and spermatozoa to the female's genitalia. The female consumes the nuptial gift and is fertilized by the spermatozoa. Females then deposit masses of eggs on plants that hang over streams. When the eggs hatch, the larvae fall or climb into the water, where they prey on the aquatic larvae of mayflies, caddisflies, and midges. Dobsonfly larvae live for up to 5 years and go through ten or twelve molts before they climb out of the water and pupate.

FISHFLIES, SPRING FISHFLIES

Latin name: *Chauliodes* sp.

Family: Corydalidae

Identification: 2- to 3-inch wingspan, gray, males have feathery antennae, similar to dobsonflies but smaller mandibles, similar to alderflies but much larger

Distribution: Eastern North America from southern Canada to Texas

Comments: These soft-bodied nocturnal insects rest on vegetation during the day. At night they fly awkwardly and are often attracted to lights near streams, ponds, and lakes, where their aquatic larvae live. Female *Chauliodes* lay their eggs on branches that overhang streams or ponds. When the eggs hatch, the larvae fall into the water. These flat predacious larvae, as well as the larvae of other insects in the family Corydalidae, are called hellgrammites by anglers, who use them for bait. They are dark brown, somewhat flattened, six-legged larvae with long filamentous gills that extend from the rear of their body. Hellgrammites also have two long retractable tubes that they can breathe through by extending them above the surface of the water. They spend 1 to 3 years as predacious larvae before they crawl out of the water, breathe through their spiracles, and pupate in the soil or rotting wood.

ALDERFLES

Latin name: *Sialis* sp.

Family: Sialidae

Identification: 0.9 to 1.1 inches long, stocky body, dark wings held rooflike over their body, lack ocelli, long filamentous antennae

Distribution: Throughout the United States and southern Canada

Comments: Alderflies are related to fishflies and dobsonflies. Adults do not feed and never fly far from the lakes, streams, and rivers where they lived as aquatic larvae, as they are poor fliers and only live for a week or so. Female alderflies lay large numbers of eggs on vegetation over lakes, streams, and rivers. When the eggs hatch, the larvae drop into the water. The larvae are predacious, with strong, sharp mandibles that they employ to catch and kill the little invertebrates on which they feed. They usually wait for their prey to pass by rather than hunt for them. Alderfly larvae breathe through seven filamentous gills on each side of their abdomen. After 1 or 2 years, they climb out of the water and pupate in underground cells that they construct along the shores of the lakes, streams, or rivers where they matured.

OWLFLIES

Subspecies: Myrmeleontiformia

Family: Ascalaphidae

Identification: 1.3 to 1.8 inches long, 2.5-inch wingspan, large eyes, similar to dragonflies but with long antennae with a knob at the end

Distribution: Southern and western United States

Comments: Owlflies are seen flying at sunset and sunrise, looking for the insects on which they prey. During the day, they rest on stems with their body, legs, and antennae pressed against the stems. Females lay their eggs near the tips of twigs, usually near streams. Their larvae are ambush predators that usually hide in debris, sometimes covering themselves with the debris for camouflage, as they wait for small insects to pass by. Larvae construct spheroidal silk cocoons when they are ready to pupate.

Male snakefly

SNAKEFLIES

Latin name: *Aguella* sp.

Family: Raphidiidae

Identification: 0.4 to 0.6 inch long, long prothorax, large eyes

Distribution: West of the Rocky Mountains from British Columbia to Arizona

Comments: Snakeflies are soft-bodied insects related to neuropterans, like fishflies and lacewings. The thorax of insects has three segments: the prothorax, mesothorax, and metathorax. In snakeflies, the prothorax is elongated. When these predators strike, they lift their head and strike somewhat like a snake, which is how they got their name. Snakefly larvae are predacious, although being soft-bodied, both adults and larvae can only take on small soft-bodied insects like aphids and mites. Females use their long ovipositor to lay eggs in crevices in the bark of conifers. Snakefly larvae can take 2, 3, or in some species up to 6 years to develop. They may go through as many as a dozen molts, although unlike most insects, there appears to be no precise number of molts.

Female snakefly

23
BUTTERFLIES

There are about 700 species of butterflies in North America. Because of the unique color patterns of their wings, many butterflies can be identified to species by comparing them to photographs. The best way to find butterflies is to look for them in places that they usually frequent. Many species of butterflies feed on nectar, so a good place to find them are places where there are lots of flowers. Planting wildflowers is a good way to attract butterflies. Also, different species of butterflies are active at different times of the spring, summer, and fall, so it is good to look for them throughout these seasons. Male butterflies of many species can sometimes be found in muddy places, feeding on salt and minerals in the mud. Another good place to find butterflies is on hilltops because male butterflies often look for mates there. Additionally, male butterflies look for females by flying back and forth along dirt roads or paths. Since the caterpillars of most species feed on a particular plant, female butterflies can often be found near those plants because that is where they lay their eggs.

Indian skipper, *Hesperia sassacus*

SKIPPERS

Subfamily: Papilionoidea

Family: Hesperiidae

Identification: 1- to 1.3-inch wingspan, tips of antennae club-shaped and usually hooked, stocky body, large eyes

Distribution: Throughout North America

Comments: Skippers are small, chubby, day-flying insects with relatively small wings. They resemble butterflies and moths, though they are more closely related to butterflies and usually considered to be a kind of butterfly. Skippers can be identified by the unique hook-shaped extension on the tips of their antennae, their stocky body, and rapid, darting flight. When perched, skippers usually hold their forewings and hindwings at slightly different angles. Although it is relatively easy to distinguish skippers from butterflies and moths, it is not so easy to identify their species. With the exception of a few species that have distinct coloration, most species look pretty much alike. Furthermore, males and females of many species have different color patterns, and colors of individuals within a species often vary. Experts distinguish species by examining their genitalia. A helpful method to distinguish species is where they are found because many species have a narrow range of only a few states, one state, or one area within a state.

Least skipper, *Ancyloxpha numitor*

ROCKY MOUNTAIN PARNASSIAN, ROCKY MOUNTAIN APOLLO

Latin name: *Parnassius smintheus*

Family: Papilionidae

Identification: 2.5- to 3.5-inch wingspan, white to yellow-brown wings with red and black markings

Distribution: Rocky Mountains from British Columbia to New Mexico, Alaska

Comments: *Parnassius smintheus*, like other butterflies of the genus *Parnassius*, live in high mountains. They are found in alpine and subalpine meadows rather than forests because they prefer light. The color of their wings varies among individuals, as the amount of melanin in their wings depends on the altitude at which they live. Butterflies that live at higher altitudes usually have more melanin because it is colder at high altitudes and the dark color helps them to warm up in the sun. Adult butterflies feed on nectar, while the larvae feed on leaves of the succulent plant *Sedum lanceolatum*. This plant is avoided by grazing mammals because it contains the poison sarmentosin. Caterpillars sequester the poison and retain it as adult butterflies. Male butterflies fly over meadows, searching for females. When they mate, the male inserts a mating plug into the female's genital opening that prevents her from mating with other males. The mating plug also contains nutrients that help her eggs develop properly.

CABBAGE WHITE BUTTERFLY, SMALL CABBAGE WHITE

Latin name: *Pieris rapae*

Family: Pieridae

Identification: 1.3- to 2-inch wingspan; horizontal black patch on wingtip; males with 1 spot, females with 2; wings light green underneath with numerous tiny dark spots

Distribution: Throughout the United States and southern Canada

Comments: Since cabbage white butterflies were imported to Quebec around 1860, they have spread throughout North America, where they live in meadows, fields, small forest clearings, and small open areas of towns, as well as gardens and farms. The adults feed on nectar from flowers, with a preference for purple, blue, and yellow flowers. Females lay eggs on members of the cabbage family. Caterpillars are bluish green with black rings around their spiracles. They feed at night and usually rest during the day under leaves, where they are less likely to be seen by predators. Cabbage white caterpillars are voracious, eating the leaves of the host plant, boring into the plant, and feeding on new sprouts. The mustard oil from the plants they eat renders them distasteful to birds. They are considered to be serious pests of members of the cabbage family, including cabbages, kale, radishes, broccoli, and horseradish.

CLOUDED SULPHUR, COMMON SULPHUR

Latin name: *Colias philodice*

Family: Pieridae

Identification: 0.8- to 1.3-inch wingspan, yellow-silver spot in center of hindwing surrounded with 2 red rings

Distribution: Throughout North America to Alaska, absent from Deep South and most of West Coast

Comments: These little butterflies often hybridize with other species of the genus *Colias*. Hybrids can be confused with purebreds. Clouded sulphurs live in alfalfa and clover fields, meadows, and roadsides. Large groups of these butterflies often congregate at mud puddles, where they feed on the mineral-rich mud. Adults feed on the nectar of a number of flowers, including butterfly bush, milkweed, coneflower, alfalfa, dandelions, and clover. Males attract mates by flying next to them and releasing a pheromone. Females that detect the pheromone lower their abdomen so that the males can mount them. Young females have difficulty distinguishing the pheromone of their species from that of other species of the genus *Colias*, which is why hybrids are common. Females lay their eggs on host plants such as soybean, sweet white and red clover, and vetch. Clouded sulphur caterpillars are green with a white stripe running along each side.

EASTERN TAILED-BLUE, EASTERN TAILED-BLUE HAIRSTREAK

Latin name: *Cupido comyntas*

Family: Lycaenidae

Identification: 0.8- to 1.4-inch wingspan, wings of males pale blue with small orange spots at base of tail, bluer in spring, wings of females gray with row of curved white spots

Distribution: Eastern United States and southern Canada to North Dakota and Texas, absent from southern Florida

Comments: The tails of hairstreaks and swallowtails have a function. The most serious predators of butterflies are birds. To avoid them, most butterflies fly low over fields, but this does not always work. If a bird attacks a hairstreak, it may mistake the butterfly's tail for its head, especially since some species of hairstreaks have a marking next to their tail that resembles an eye. The bird then ends up with the tail and maybe a bit of wing, and the butterfly escapes with its life. Missing a bit of wing does not seriously affect the ability of butterflies to fly. The caterpillars of these butterflies have an unusual symbolic relationship with ants. Caterpillars feed on various legumes, where they secrete a substance on which ants feed. The ants, in turn, protect the caterpillars from predators.

Newly emerged black swallowtail

BLACK SWALLOWTAIL, AMERICAN SWALLOWTAIL

Latin name: *Papilio polyxenes*

Family: Papilionidae

Identification: 3.3- to 4.3-inch wingspan, females dusty black with blue markings, males black with yellow markings

Distribution: Southern Canada and the United States east of the Rocky Mountains, south to Mexico

Comments: Female black swallowtails lay their eggs on members of the parsley family, and the caterpillars absorb toxins from the plant that make them toxic to birds. When they are ready to pupate, the caterpillars defecate for the last time and leave their food plant, as predators often find caterpillars by the scent of their frass. Caterpillars of all stages of these and other swallowtail butterflies have a structure called the osmeterium behind their head, which is usually withdrawn.

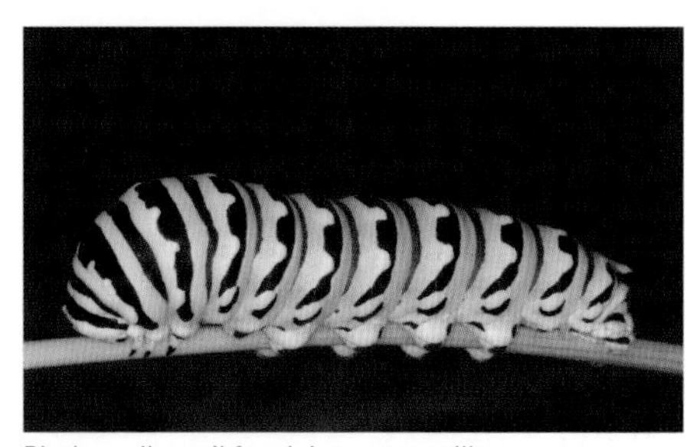

Black swallowtail fourth instar caterpillar

When it senses danger, the caterpillar extends the organ, which looks like a snake's tongue and emits a fouls odor that repels predators. Swallowtails overwinter as pupae (chrysalises), which can be green or brown. Adult butterflies feed on the flowers of a number of plants, including clover and milkweed. Black swallowtails mimic another swallowtail, the pipevine swallowtail (*Battus philenor*), which is poisonous.

Spicebush swallowtail caterpillar

SPICEBUSH SWALLOWTAIL

Latin name: *Papilio troilus*

Family: Papilionidae

Identification: 3- to 4-inch wingspan, black with ivory-colored spots on wings, orange spots on upper surface of hindwings

Distribution: Eastern United States and southern Ontario, occasionally in Colorado and Manitoba

Comments: Spicebush swallowtails are named for the food plant of their caterpillars, although the caterpillars may also feed on a number of other plants including white sassafras, prickly ash, tulip tree, and sweet bay. The front legs of female spicebush butterflies have chemoreceptors. By touching their forelegs to a leaf, they can determine if the plant is a suitable place for them to lay their eggs. When a spicebush butterfly egg hatches, the caterpillar chews the leaf from the edge to the middle vein while it releases silk. When the silk dries, it causes the leaf to fold around the caterpillar, creating a tube for the caterpillar to hide during the day. Caterpillars come out of the tube to feed at night. The first instar caterpillars are brown and mimic bird droppings. However, when they reach the fourth and final instar, they have markings that resemble the eyes and nose of a snake, which may frighten a bird.

BALTIMORE CHECKERSPOT

Latin name: *Euphydryas phaeton*

Family: Nymphalidae

Identification: 2- to 3-inch wingspan, black with orange and white spots

Distribution: Maine, south to South Carolina, west to Minnesota and Oklahoma, southern Quebec and Ontario

Comments: The Baltimore checkerspot is the state insect of Maryland. It is named after George Calvert, the first Lord Baltimore, whose crest was black and orange. Females lay a few hundred eggs under the leaves of their host plant, the turtlehead (*Chelone glabra*). When the eggs hatch, the larvae move to the tips of the plant, where they build communal webs and feed together. In the late summer or fall, the fourth instar caterpillars crawl to the ground, where they overwinter. In the spring they feed on a number of plant species before they pupate. Overwintering as caterpillars is unusual for butterflies, as most species over-winter as pupae or eggs. Adults feed on nectar from the flowers of milkweed and viburnum.

Birds often bite off pieces of the wings of butterflies. Missing a bit of wing does not appear to affect the butterfly's ability to fly.

PEARL CRESCENT

Latin name: *Phyciodes tharos*

Family: Nymphalidae

Identification: 1.3- to 1.7-inch wingspan, orange wings with black markings, females darker than males

Distribution: Throughout eastern North America, west to the Rocky Mountains

Comments: Pearl crescents are very daring butterflies. They often fly after passing insects, including butterflies of different species, and even birds and Frisbees—yes, the toy. Females lay about 700 to 800 eggs in clusters of 20 to 400 on the underside of leaves of plants in the aster family. Caterpillars usually feed together. When they are ready to pupate, the caterpillars crawl some distance from the host plant and suspend themselves by a holdfast as they pupate. Usually pearl crescents have three generations in a season. They overwinter as third instar caterpillars.

MONARCH BUTTERFLY, MONARCH

Latin name: *Danaus plexippus*

Family: Nymphalidae

Identification: 3.5- to 4-inch wingspan; wings have distinctive black, orange, and white pattern; black head and thorax with white spots; black legs

Distribution: Throughout the United States and southern Canada

Comments: Monarch butterflies are well known for their annual migration south from the northern United States and southern Canada to Florida and Mexico. During the fall the butterflies cover thousands of miles and return in the spring. Individuals do not live long enough to make the entire trip. Instead, they produce new generations that continue the journey. Eventually, some butterflies reach their principal destination, the Mariposa Monarca Biosphere Reserve in Mexico, where many thousands of monarchs gather in trees and bushes.

Monarch butterfly caterpillar

Monarchs lay their eggs on milkweed, which is a poisonous plant. When the eggs hatch, the caterpillars feed on the leaves of the plant before they pupate. They sequester the poison, to which they are immune, and retain it when they pupate into adults. Farmers often use herbicides to destroy milkweed plants because they can poison livestock that feeds on the plant. This practice, along with the widespread use of pesticides, has diminished monarch populations.

BUCKEYE, COMMODORE, PANSY

Latin name: *Junonia coenia*

Family: Nymphalidae

Identification: 2- to 2.2-inch wingspan; olive brown with white, black, and orange eyespots

Distribution: Throughout North America

Comments: Due to genetic differences and seasonal variations, the pattern of eyespots is variable within a species. These migratory butterflies are strong fliers. They can be found in southern Canada in the summer, but they fly south in the fall. Usually, there are two generations a year. Caterpillars are stout, dark green to black, with blue-black branched spines. They feed on many plants including Labiatae, Acanthaceae, Amaranthaceae, and Scrophulariaceae.

RED ADMIRAL

Latin name: *Vanessa atalanta rubria*

Family: Nymphalidae

Identification: 2-inch wingspan, black wings, white spots and orange bands

Distribution: Throughout North America

Comments: Although the monarch butterfly is by far the most well-known butterfly species that migrates, there are a number of other species of butterflies, as well as moths, that migrate. The red admiral is one of them. These butterflies cannot survive the winter in the north, so in the fall they fly south, where they lay their eggs and die. In the spring, the next generation heads north. The population of these butterflies in the north varies from summer to summer. Presumably, the number of red admirals that return depends on the weather conditions along their migration path to and from the south, as well as the conditions in the south during the winter. Butterfly migration is hazardous.

Red admiral underwings

AMERICAN LADY CATERPILLAR

Latin name: *Vanessa virginiensis*

Family: Nymphalidae

Identification: Caterpillars black with yellow stripes, red dots, and multiple spikes

Distribution: Throughout the United States and southwest Canada

Comments: As these caterpillars are mostly diurnal and must feed constantly, they are continually exposed to predators, especially birds. Furthermore, because caterpillars can't fly and are very slow-moving insects, they have no way of escaping. The American lady caterpillar uses silk to sew together bits of leaves to make a nest to hide in. Also, like certain other caterpillars, American lady larvae are covered with sharp spikes to discourage birds from attacking them and to render them distasteful. They also curl up into a ball to expose their spikes so as to appear even more unappetizing. Birds usually avoid caterpillars that are adorned with spikes. When the caterpillar pupates, the pupa is green in color to match the shade of the plant on which it pupates. Not only does the pupa have colors to match its surroundings, but it is also shiny if it pupates on a shiny plant and dull if the plant is dull.

ZEBRA BUTTERFLY, ZEBRA LONG-WING, ZEBRA HELICONIAN

Latin name: *Heliconius charitonius*

Family: Nymphalidae, subfamily Heliconiinae

Identification: 2.8- to 3.5-wingspan, long-winged, black with distinctive white stripes

Distribution: Florida and along the Gulf Coast, occasionally migrating to Georgia, South Carolina, and Arizona in summer

Comments: Adults feed on pollen and nectar. They fly slowly with shivering wingbeats through open woodlands and along borders of tropical hemlocks. Zebra butterflies roost communally at night in groups of up to several dozen. They synthesize cyanogenic glycosides, which are poisons that render them toxic to predators. Birds recognize the zebra butterfly's pattern and usually do not attack them. A number of unrelated butterflies that are not poisonous mimic the pattern on zebra butterflies. Unfortunately, mass spraying of the organophosphate insecticide, Naled, for mosquitoes in Miami-Dade County, Florida, has killed most of these beautiful butterflies.

24
MOTHS

Most of the approximately 10,000 species of North American moths are nocturnal. Moths can usually be distinguished from butterflies by their antennae. The antennae of most female moths are long and thin, and those of males are featherlike. On the other hand, the antennae of butterflies have a knob at the end. The good news is that moths are attracted to lights, so they will come to you. If you live in an apartment, you may have to venture out in the evening to a lighted building like a store that has a big window. Although moths are especially attracted to ultraviolet light, incandescent light is sufficient. Moths are not the only insects that are attracted to lights. In fact, many night-flying and some diurnal insects come to lights, so wooded areas and fields near lights will attract many insect species. The best time to look for moths and other insects is on hot, humid evenings. Another way to attract moths is to paint some watered-down molasses on a tree. Instead of molasses, some people make various concoctions of sugar, fruit juice, bananas, and even beer to attract moths.

AILANTHUS WEBWORM

Latin name: *Atteva punctella*

Family: Attevidae

Identification: 0.7- to 1.3-inch wingspan, bright orange forewings with 4 black bands with white spots

Distribution: Great Lakes, south to Texas, east to New York and Florida

Comments: Ailanthus webworms are exceptionally beautiful diurnal moths with a distinctive orange, black, and white pattern on their forewings. They are found across North America but are believed to have originated in Florida or in Central or South America, where the caterpillars fed on the paradise tree (*Simarouba glauca*). However, when a Chinese tree, the tree-of-heaven (*Ailanthus altissima*), became popular in North America, the moth switched the food plant of the caterpillars to *Ailanthus*, which is closely related to the paradise tree. Unlike most moths that spread their wings out when they are at rest, ailanthus webworms tuck in their wings. Once absent from the north, these beautiful moths can now be found over most of northern North America, although in the fall they fly south for the winter. The moths are called webworms because their caterpillars shelter for protection in communal silk webs that they construct in the *Ailanthus* tree.

Ailanthus webworm feeding on nectar

AZALEA SPHINX MOTH, AZALEA HAWK MOTH

Latin name: *Darapsa choerilus*

Family: Sphingidae

Identification: 2- to 3-inch wingspan, light orange-brown wings and body with horizontal white stripes

Distribution: Southeastern Canada to Florida, west to the Mississippi Valley

Comments: Sphinx moths have long tongues that suck deep-seated nectar in tubular flowers. Female azalea sphinx moths extend a scent gland from the posterior of their abdomen to attract males. After mating, eggs are laid on azalea or viburnum bushes. Caterpillars feed on the plant until they are ready to pupate. They then crawl to the ground, dig a hole, and pupate. If they are the last brood of the season, they overwinter as pupae. There are usually two broods in the north and as many as six or eight in the Deep South. Like other sphinx moths, the small azalea sphinx moth is fast and agile in flight. The dynamic flight of these and other hawk moths is facilitated by their antennae that vibrate in a plane such that the moth is able to detect the position of its body relative to the surface of the Earth.

HOG SPHINX, VIRGINIA CREEPER SPHINX, GRAPEVINE SPHINX

Latin name: *Darapsa myron*

Family: Sphingidae

Identification: 1.7- to 2.6-inch wingspan, forewings usually brown and tan or yellowish gray but can be greenish, underside of hindwings pale orange

Distribution: Maine to Florida, west to eastern Texas, also southern Ontario

Comments: Like other sphinx moths, hog moths feed on nectar. The host plants of their night-feeding caterpillars are grape, Virginia creeper, and pepper vines. Because these and other sphinx moths have an especially long proboscis, they are able to obtain nectar from bell-shaped flowers, where the nectar is deep within the flower. Sphinx moths have the ability to hover over flowers, which enables them to visit more flowers than they would be able to if they had to land on each one. Hovering also helps sphinx moths to escape predators because they are already airborne. Also, they can often evade birds or bats because, being smaller, they can outmaneuver them. Small flying animals can usually turn faster than larger ones because they have less momentum, just as small airplanes or boats can turn faster than larger ones.

SMALL-EYED SPHINX MOTH, SPHINX MOTH, HAWK MOTH

Latin name: *Paonias myops*

Family: Sphingidae

Identification: 1.7- to 3-inch wingspan, forewings smoothly indented, hindwings have small yellow patches enclosed in a single eyespot

Distribution: Primarily in the United States and Canada, east of the Mississippi, occasionally in western states and southwestern Canadian provinces

Comments: These moths get their name from the single black and blue eyespot on their hindwings. Like other species of sphinx moths, the small-eyed sphinx moth flies fast and can hover well. Their hairless green caterpillars have a spinelike horn on their back, which is why they, as well as the caterpillars of hawk moths, are called hornworms. When they are resting, caterpillars usually hold their legs off the surface and tuck their head underneath their legs, giving them the appearance of a sphinx, which is why the adults are called sphinx moths. Small-eyed sphinx moth caterpillars feed on a number of plants, but their preferred plant is cherry. There are two generations of these sphinx moths a year in the northern United States and Canada and up to four in the southern states.

HUMMINGBIRD MOTH

Latin name: *Hemaris thysbe*

Family: Sphingidae

Identification: 1.5- to 2-inch wingspan, wings without scales, forewings larger than hindwings, bumblebee and hummingbird mimic

Distribution: Across the United States, southern Canada, northwest to the Yukon and western Alaska

Comments: Hummingbird moths are members of the hawk moth family. Unlike most sphinx moths, hummingbird moths are diurnal, although they sometimes fly at dusk. Just like hummingbirds, these moths hover over flowers and can fly backward. In fact, if you are not familiar with these moths, you might easily mistake one for a hummingbird. You might also mistake one for a bumblebee, although they are larger and bumblebees usually land on flowers, while hummingbird moths hover when they feed. These moths prefer to feed on butterfly bushes and honeysuckle, but they also feed on the nectar of other flowering plants. Caterpillars feed on the leaves of honeysuckle, viburnum, dogbane, hawthorn, cherry, and plum. When they are ready to pupate, the caterpillars fall to the ground, spin a cocoon, and pupate in the leaf litter.

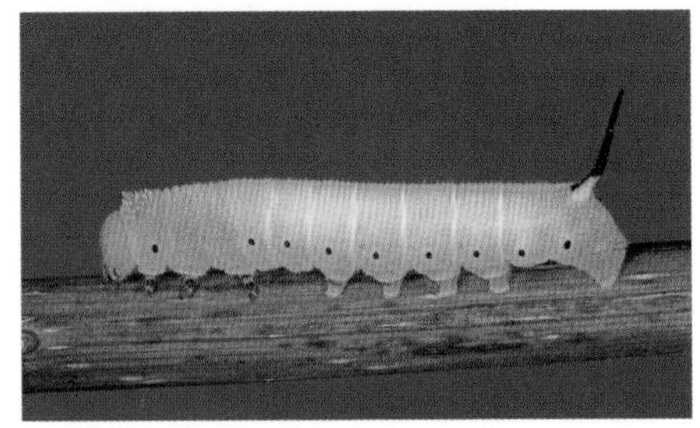

Hummingbird moth caterpillar (hornworm)

SNOWBERRY CLEARWING

Latin name: *Hemaris diffinis*

Family: Sphingidae

Identification: 1.2- to 2-inch wingspan, golden thorax, black abdomen with golden posterior segments, black legs, wings mostly clear with brown terminal borders

Distribution: Eastern North America and southern Canada to the Rocky Mountains

Comments: Snowberry clearwings are day-flying moths. Their gold and black coloration makes these moths convincing bumblebee mimics. In fact, they often are seen in the company of bumblebees as they hover and feed on the nectar of flowers of lantana, orange hawkweed, thistle, lilac, violet, snowberry, and other plants in the honeysuckle family. Those that live in the north have one generation a year, while those in the south have two or more. After they mate, the females lay eggs on the undersides of leaves of one of their larvae food plants, which include honeysuckle, dogbane, dwarf bush honeysuckle, and snowberry. When their green caterpillars are ready to pupate, they drop to the ground and form a loose cocoon in the leaf litter, where they overwinter and emerge as adults in the spring.

BANDED TUSSOCK MOTH, PALE TIGER MOTH

Latin name: *Halysidota tessellaris*

Family: Erebidae

Identification: 0.8 to 1 inch long, pale tan thorax with turquoise and yellow stripes

Distribution: Eastern United States, west to Minnesota and Texas, southern Quebec and Ontario

Comments: Banded tussock moth caterpillars feed on a number of deciduous trees and shrubs, including alder, ash, birch, elm, oak, and willow. However, they are never numerous enough to cause serious damage to the trees. Adults are attracted to decaying plants that contain pyrrolizidine alkaloids, poisonous compounds that many plants synthesize to defend against insects that feed on them. Banded tussock moths regurgitate a fluid substance onto the leaves of these plants. They then acquire the poisonous pyrrolizidine alkaloids by drinking the regurgitated fluid off the leaves.

Male gypsy moth

GYPSY MOTH

Latin Name: *Lymantria dispar*

Family: Erebidae

Identification: 2.2- to 2.6-inch wingspan, males straw-colored with brown markings, females cream-colored

Distribution: Southern Canada and northern United States

Comments: Gypsy moths were brought to Bedford, Massachusetts, from Europe in 1890 by a Frenchman named Trouvelot, who had the naive idea of crossing the gypsy moth with the silk moth. He believed that the hybrid would be easier to raise because gypsy moth caterpillars feed on the leaves of many different kinds of trees, whereas silk moth caterpillars feed only on the leaves of the mulberry tree. However, the two species are not closely related and thus do not interbreed. Unfortunately, some of Trouvelot's gypsy moths escaped and have gradually expanded their range. Adult moths have no digestive system and do not feed. On the other hand, the caterpillars feed on the leaves of numerous species of deciduous trees, thereby weakening the trees. Male gypsy moths are attracted to lights. Although female gypsy moths have large wings, they can't fly. The females secrete sex pheromones to attract mates. It is easiest for a male to find a female if she stays put, which is why females cannot fly.

Female gypsy moth

Isabella tiger moth caterpillar, or woolly bear

ISABELLA TIGER MOTH, WOOLLY BEAR MOTH

Latin name: *Pyrrharctia isabella*

Family: Erebidae

Identification: 1.6- to 2.2-inch wingspan, light brown with small black spots on forewings, caterpillars have black hair on anterior and posterior and brown hair in center

Distribution: Throughout the United States and southern Canada

Comments: It is said that one can predict how cold the winter will be by the relative lengths of the brown and black areas of the woolly bear caterpillar's body. This prediction is as reliable as predicting whether it will be an early spring or 6 more weeks of winter depending on whether a groundhog sees its shadow. Woolly bears synthesize a poison and cover their hairs with the poison so that they will not be palatable to predators. Birds usually recognize woolly bears and stay away from them. When attacked, caterpillars curl themselves into a ball to protect their delicate ventral side. In the northern part of their range, the summer doesn't last very long, so woolly bear caterpillars must spend several seasons as caterpillars before they can pupate. When they form their cocoon, woolly bears remove their poisonous hairs one by one and weave them into the cocoon.

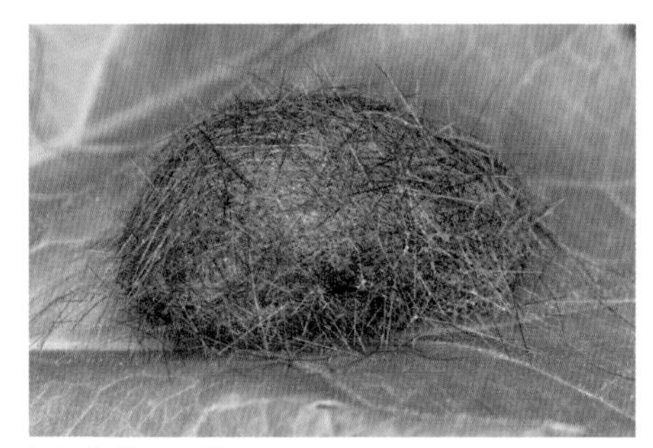

Isabella tiger moth cocoon

UNDERWING MOTHS, UNDERWINGS

Latin name: *Catocala* sp.

Family: Erebidae

Identification: 1.2- to 1.8-inch wingspan, most species mottled gray or brown, top of hindwings have bright black and orange, red, white, or blue stripes

Distribution: Throughout North America

Comments: There are over a hundred species of the genus *Catocala* in North America. Some species have colorful common names like darling underwings, girlfriend underwings, and sweetheart underwings. Although they are called underwings, these moths are named for the bright colors on the upper side of their hindwings. The forewings of these nocturnal moths are camouflage-colored to blend in with the bark of the trees where they rest during the day. However, when they are disturbed, they move their forewings forward to reveal their brightly colored hindwings and fly away. It is believed that the sudden flash of their bright hindwings serves to startle predators, giving the moths enough time to escape.

VIRGIN TIGER MOTH

Latin name: *Grammia virgo*

Family: Erebidae

Identification: 2.2-inch wingspan, black forewings with white markings, orange hindwings with black spots

Distribution: Maine to South Carolina, west to the Mississippi River and Alberta

Comments: When they are at rest, the beautiful orange hindwings of virgin tiger moths lie beneath their black and white forewings. However, when these moths feel threatened, they move their forewings toward their head to reveal their brightly colored hindwings to startle and warn a bird or other predator. The orange and black warning colors also remind birds that virgin tiger moths are poisonous. When attacked, adults can secrete blood containing toxic alkaloids from their thorax. The alkaloids, which they synthesized when they were caterpillars, are retained in the adult moths. When virgin tiger moths mate, males transmit their toxic alkaloids to females along with sperm. The alkaloids coat the eggs as they are laid, protecting them from predators. Although many nocturnal moths can detect the high-pitched sounds that are made by bats, virgin tiger moths also have tympanum organs that produce an ultrasonic sound that can startle and confuse bats.

YELLOW-COLLARED SCAPE MOTH

Latin name: *Cisseps fulvicollis*

Family: Erebidae

Identification: 0.8- to 1.2-inch wingspan, slate black with bright orange thorax

Distribution: Across the United States and southern Canada

Comments: Yellow-collared scape moths have a bright orange (not yellow) thorax. When you see an orange insect, or an insect with orange coloring somewhere on its body, you can be almost certain that it is either poisonous or mimicking a poisonous insect. In this case, male moths feed on plants of the Eupatorium family, commonly called thoroughworts, snakeroots, or bonesets. These plants contain poisonous alkaloids, but the yellow-colored scape moth is immune to the poison. The poison that protects the male moths is transferred to females when they mate. As the eggs pass down the female's oviduct, they become coated with the alkaloids that protect them from predators.

CECROPIA MOTH

Latin name: *Hyalophora cecropia*

Family: Saturniidae

Identification: 5- to 6-inch wingspan, distinctive crescent-shaped pattern on wings

Distribution: Across North America and southern Canada

Comments: The cecropia is the largest moth in North America. Adults have only vestigial mouthparts, do not feed, and only live for a few weeks. Males can detect female sex pheromones from a distance of several miles. Unfortunately, these exquisite insects have many predators. They overwinter as pupae that are often eaten by squirrels. However, the most threatening is a parasitoid fly, *Compsilura concinnata*, which was introduced to North America from Europe to control the gypsy moth but is also a parasitoid of cecropia and other caterpillars.

IO MOTH

Latin name: *Automeris io*

Family: Saturniidae

Identification: 2.5- to 3.5-inch wingspan, males have yellow forewings, females have brown forewings, hindwings of both sexes have a prominent blue eyespot outlined in black

Distribution: Eastern United States, west to Montana and New Mexico; southern Canada

Comments: The large black and blue eyespots on the hindwings of the io make these beautiful large moths one of our most recognizable species. These eyespots are meant to frighten would-be predators. A number of parasitoid wasps lay their eggs in moth eggs. Each io moth egg has a black spot. When a parasitoid wasp sees the spot, she may think that it is where another wasp had made a hole when she laid her eggs and will go on her way. Io moths are nocturnal. In the daytime, they hide in the dead brown-red leaves that are common on the forest floor. Mature caterpillars of io moths are green with a brown and white stripe on their sides and numerous spines on their back. It is best to leave these caterpillars alone, as the spines are poisonous and often break off in your skin and can be irritating for hours.

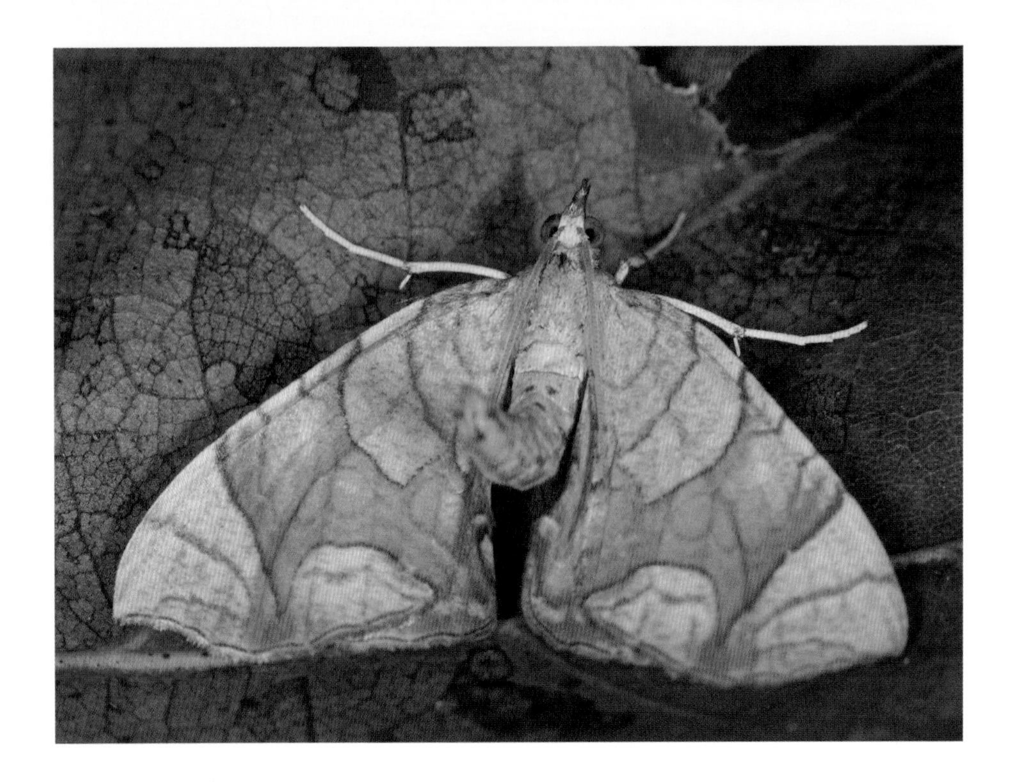

LESSER GRAPEVINE LOOPER

Latin name: *Eulithis diversilineata*

Family: Geometridae

Identification: 1.5- to 2-inch wingspan, straw-colored with thin dark brown stripes

Distribution: Primarily in eastern United States and southern Quebec and Ontario

Comments: When these nocturnal moths rest, they curl their abdomen up and forward and hold their wings flat to reduce their shadow so they are less likely to be noticed by birds and other predators. "Looper" refers to the caterpillars of these moths and other Geometrid moths. They are also referred to as inchworms or spanworms. Some, like the lesser grapevine looper, are often called twigworms, because when they rest, they grasp a twig with their prolegs and extend their body in the air to disguise themselves as twigs. Most caterpillars have three pairs of legs and five pairs of prolegs, but loopers have only one or two pairs of prolegs. When they walk, they move their body forward, grab on to a branch with their front legs, and then let go with their prolegs and move the back part of their body to their forelegs to form a loop. Caterpillars feed on grape leaves and Virginia creeper.

CROCUS GEOMETER MOTH, FALSE CROCUS GEOMETER

Latin name: *Xanthotype urticaria*

Family: Geometridae

Identification: 1.2- to 1.6-inch wingspan, yellow with variable patches of blotched brown

Distribution: Northeastern United States and southeastern Canada

Comments: The crocus geometer moth is in the family Geometridae. The name of the family derives from *geometer*, which means "earth measurer." It refers to the fact that they are loopers, or inchworms, and each time they take a step, it is 1 inch in length. Usually, they are green or brown and depend on their camouflage to avoid the many insects and other animals that prey on caterpillars. The caterpillars of crocus geometer moths feed on dogwood, ground ivy, rhododendron, and goldenrod. There are five species of *Xanthotype* with overlapping ranges. It is not possible to tell them apart from pictures because their wing patterns are so variable.

CURRANT SPANWORM

Latin name: *Itame ribearia*

Family: Geometridae

Identification: 1- to 1.25-inch wingspan, adult moth white or tan with variable yellow and brown markings, caterpillar white with black and yellow pattern

Distribution: Uncommon, throughout northeastern United States

Comments: The currant spanworm is a rather dull, light brown moth, but the caterpillars are beautiful. Many other kinds of dull-looking moths have beautiful and interesting caterpillars. As the name implies, the caterpillars feed on currants. In addition to currants that are farmed, there are about 150 known species of wild currants, genus *Ribes*. Unfortunately for wild currants and for the currant spanworm, currant bushes are a secondary host for white pine blister rust, an Asian fungus that damages white pine trees. The fungus was introduced to North America around the turn of the twentieth century. To mitigate the fungus, in the 1920s the United States Department of Agriculture began a program to eradicate wild currant bushes in the United States. The program lasted for 40 years. Currant spanworms are not very common because they were collateral victims of the currant eradication program.

RED-FRINGED EMERALD CATERPILLAR

Latin name: *Nemoria bistriaria*

Family: Geometridae

Identification: Strange-looking larvae with brown with white speckles, sawtooth back, 2 pairs of prolegs

Distribution: Eastern North America, west to Texas

Comments: Adult red-fringed emerald moths are greenish, although their color varies with their diet. Diet also seems to affect the behavior of these moths. Those that emerge in the spring tend to rest on the flowers of certain trees, whereas those that emerge later in the summer rest on stems and twigs. Their strange-looking caterpillars feed on the leaves of white oak and black walnut trees. Their unusual appearance may protect them, because birds may not recognize their bizarre form as that of a caterpillar.

PALE BEAUTY

Latin name: *Campaea perlata*

Family: Geometridae

Identification: 1.2- to 2-inch wingspan, pale green or white wings crossed with 2 light brown and white transverse lines

Distribution: Throughout North America including Alaska, more common in the Northeast

Comments: These pale green or white nocturnal moths are common across North American from May until September. Their caterpillars feed on several species of conifers, as well as deciduous trees and shrubs including alder, ash, basswood, beech, birch, blueberry, cherry, elm, maple, and oak. They are often called finger loopers because they have short hairs like fingers along their sides that when pressed against a branch make the outline of the caterpillar almost invisible. Pale beauties overwinter as third or fourth instar caterpillars.

Smaller flightless female on left, male on right

WINTER MOTH

Latin name: *Operophtera brumata*

Family: Geometridae

Identification: Males 0.3- to 0.4-inch wingspan, wings light gray-brown or yellow-brown; females 0.1 to 0.2 inch long, dark brown, small wings

Distribution: New England, Oregon, and Washington State

Comments: This imported moth was introduced into Nova Scotia in the 1930s. They were discovered in Massachusetts around 2013 and have been defoliating trees since then. Male winter moths usually fly after the first frost. The small, dark, flightless females rest on trees and release pheromones to attract mates. After they mate in the late fall or early winter, females deposit 150 to 350 eggs in the crevices of bark or on lichen. The adults then die. In March, the caterpillars crawl up trees and munch away at the leaves. The caterpillars feed on the leaves of a number of different trees and sometimes are so numerous that they completely defoliate the trees. Young green caterpillars often hang by their mouths from silk threads that break off in the wind and can carry them long distances. Because their mothers cannot fly, this method of travel, called ballooning, is usually how winter moths find new territories.

Male winter moth

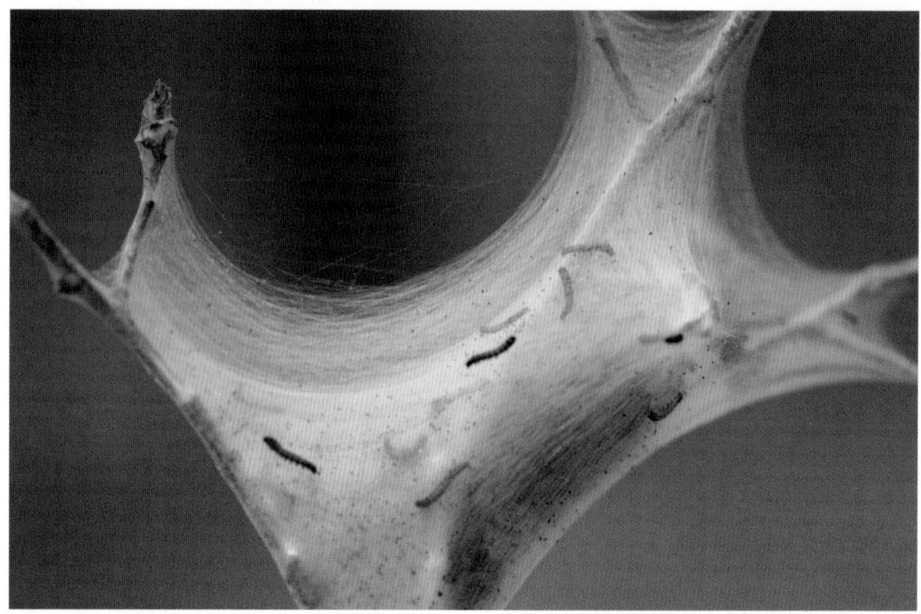

Eastern tent caterpillar nest

EASTERN TENT CATERPILLAR, LAPPED MOTH

Latin name: *Malacosoma americanum*

Family: Lasiocampidae

Identification: 0.8 to 1 inch long, light brown with 2 light stripes on forewings

Distribution: Throughout eastern United States and southern Quebec and Ontario

Comments: Female eastern tent caterpillars lay clusters of 100 to 400 eggs on cherry, apple, peach, or hawthorn trees. The caterpillars mature, but they do not leave their eggs. Instead, they remain in the egg, overwinter, and crawl from the egg the following season and begin to spin colonial nests between tree branches. Although the caterpillars spend their nights in the nest, they leave to feed several times a day. They usually crawl in a large group, leaving a silk trail along the branches of the tree, which allows them to find their way back to their nest. When they are ready to pupate, the caterpillars leave the nest and pupate individually some distance from the nest. When they detect a predator, eastern tent caterpillars thrash the anterior part of their body. This may frighten off a predator or create a moving target, making it difficult for parasitoid flies and wasps to lay their eggs on or in the caterpillars.

GRAPE LEAF FOLDER

Latin name: *Desmia funeralis*

Family: Crambidae

Identification: 1- to 1.2-inch wingspan, black with white markings

Distribution: Throughout North America

Comments: Grape leaf folder moths fly during the day from May until September. There are two or three generations per year. Female moths lay their eggs on the leaves of grapevines, redbud trees, or primrose plants. When they hatch, the caterpillars fold over the leaves of the plant. They then attach the two sides of the leaf together with silk. The caterpillars mature inside the folded leaf as they eat the leaf, grow, molt, and pupate. Several insects attack these caterpillars, including a tiny parasitoid wasp, *Bracon cushmani*, that paralyzes them with its sting and lays one or more of its own eggs in the caterpillar. When they hatch, the wasp larvae feed on the caterpillar and eventually pupate next to it.

SNOWY UROLA, SNOUT-MOTH

Latin name: *Urola nivalis*

Family: Crambidae

Identification: 0.4- to 0.5-inch wingspan, satin-colored with brown and black markings on wingtips, dark front legs

Distribution: Northeastern United States, southeastern Canada

Comments: The snowy urola is a nocturnal moth. They are attracted to lights and can sometimes be flushed from the grassy fields where they hide in the daytime. Most moths are camouflaged to look like tree bark, and they rest on trees in the daytime with their wings open and flat against the tree. However, moths of the family Crambidae spend the day hiding in grass and are sometimes called grass moths or grass-veneers. These moths rest on blades of grass with their wings held tight against their body, forming a tubular shape. Resting flat against a blade of grass, they are difficult for predators to see. Most species have long hairy palpi, which make them look like they have a big snout; therefore, they are sometimes referred to as snout-moths.

ORANGE-PATCHED SMOKY MOTH

Latin name: *Pyromorpha dimidiata*

Family: Zygaenidae

Identification: 0.7- to 1.2-inch wingspan; black and orange wings, held horizontal over abdomen; resembles net-winged beetles

Distribution: New York to Florida, west to Oklahoma and Missouri

Comments: The orange-patched smoky moth is a small black and orange day-flying moth that is common east of the Mississippi River. Their caterpillars are skeletonizers of soft plant tissue. These moths mimic the orange and black poisonous net-winged beetle *Caenia dimidiata,* with a similar color pattern on their wings and body warning predators to stay away. They frequently hide among the net-winged beetles, where birds and other predators may mistake them for the poisonous beetles. However, these moths are themselves poisonous because they synthesize hydrogen cyanide.

The net winged beetle, *Caenia dimidiata*

GRAPELEAF SKELETONIZER

Latin name: *Harrisina americana*

Family: Zygaenidae

Identification: 0.4- to 0.6-inch-long body, black with yellow or orange collar, hindwings reduced

Distribution: Eastern North America, southern Ontario, other species in western states

Comments: The orange color is a warning to predators, as these moths produce hydrogen cyanide. The abdomen is curled upward and has a fan shape. Female grapeleaf skeletonizers lay their eggs on the leaves of wild and cultivated grapes and Virginia creeper (*Parthenocissus quinquefolia*). Young larvae have yellow and hairy black stripes. They are skeletonizers, feeding on the soft parts of the grape or Virginia creeper leaf and leaving the veins untouched. The veins are used for support as the caterpillars feed.

Grapeleaf skeletonizer caterpillar

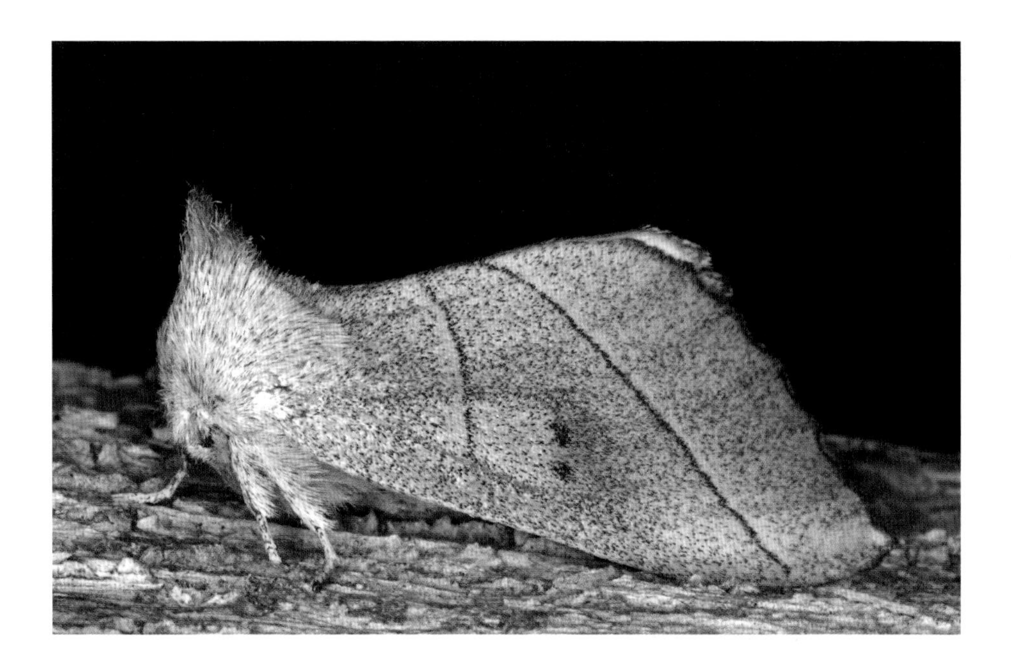

WHITE-DOTTED PROMINENT

Latin name: *Nadata gibbosa*

Family: Notodontidae

Identification: 1- to 2-inch wingspan, heavy-bodied, long wings folded across back

Distribution: Across North America

Comments: The white-dotted prominent, as well as other prominents (noctuids), have a hearing tympanum that enables them to detect the high-pitched sounds that bats make to find their way and detect flying insects. When they detect a bat, moths that are perched stay put, and those that are flying usually fall to the ground. The ability to detect sound is also employed by prominent moths to find mates. Females increase their flapping frequency when searching for mates, and males make a trembling sound in response. Nocturnal moths are attracted to lights. If you live in an apartment in a city where there is no place that you can access moths on lights, find a store where the lights are left on in the evening. I have collected many moths in this way. A few times I have been confronted by the police, but they always went away when I explained what I was doing. People who collect insects are often considered peculiar but harmless.

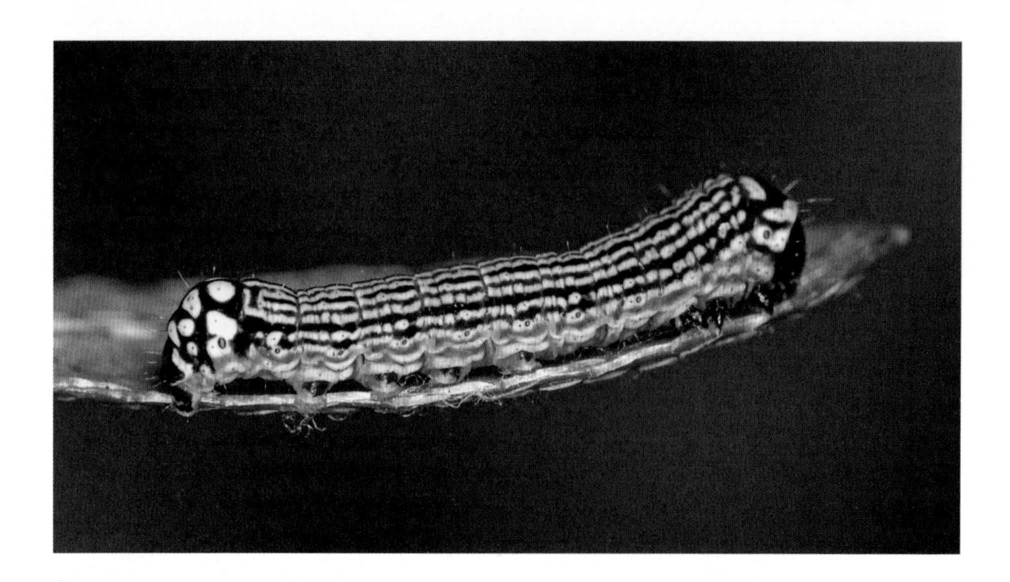

TURBULENT PHOSPHILA

Latin name: *Phosphila turbulenta*

Family: Noctuidae

Identification: 0.6 to 0.75 inch long, adult molted brown wings, caterpillar black and white stripes, underbelly yellow, false head

Distribution: Southern Maine to Florida and Texas

Comments: These little brown moths range all over the Northeast. Only an expert can distinguish them from the many other similar brown moths in the same range because the wing colors and patterns vary somewhat between individuals. However, it is easy to identify the caterpillars because they look like they have two heads. The real head on the front looks like the fake head on the back. The place to find these caterpillars is on the common greenbrier vines, *Smilax* sp., the only plant on which they feed. It is not unusual for a caterpillar to have just one or a few related food plants. This also means a female moth must seek out that plant to lay her eggs. To accomplish this, moths and butterflies employ chemoreceptors on their feet to detect the surface molecules on leaves to identify plants.

NO COMMON NAME

Latin name: *Schizura* sp.

Family: Notodontidae

Identification: Gray and green caterpillars with intricate black pattern

Distribution: Northeastern United States and southeastern Canadian provinces

Comments: Almost all moths and caterpillars are camouflaged, but their colors are quite different because of their different habitats. Most moths are nocturnal. They rest on trees during the day and are colored to match the dull, mottled colors of tree bark. On the other hand, caterpillars can be either diurnal or nocturnal depending on the species, and their color and pattern usually allow them to be camouflaged to fit in with the color of their food plant. Although all but the tiniest insects have spiracles, they usually are not visible. However, in some caterpillars, like those of the notodontid genus *Schizura*, the spiracles are very obvious because they are colored. In the case of *Schizura* sp., the caterpillar's orange spiracles are surrounded by black and white circles.

SILK MOTH, SILKWORM

Latin name: *Bombyx mori*

Family: Bombycidae

Identification: 1.2- to 2-inch wingspan, moths white, caterpillars gray with white markings

Distribution: Extinct in the wild

Comments: Silkworm is the name for the caterpillar of the silk moth *Bombyx mori*. These caterpillars produce almost all of the commercial silk in the world. Although they no longer exist in the wild, silkworms are raised in North America and sold as food for pet reptiles and amphibians. *Bombyx mori* have been domesticated for thousands of years. These moths have wings but cannot fly and have no fear of man. Growing the moths for silk, known as sericulture, is a cottage industry throughout Asia. It has never been successfully introduced into western countries because it is extremely labor-intensive, as silk must be carefully unwound from the silk moth cocoons one at a time and rewound onto a spool. Although there is an artificial diet, the natural food of the silkworm is the leaves of the mulberry tree. In Asia, wives and children on rural farms often work in sericulture while the husbands are out in the field.

Silk moth caterpillar (silkworm)

JOCOSE SALLOW, THE JOKER

Latin name: *Feralia jocosa*

Family: Noctuidae

Identification: 1.1- to 1.2-inch wingspan, olive-green forewings with black and white lines and spots, light brown hindwings

Distribution: Northeastern United States, south to Maryland and Ohio; southeastern Canada

Comments: Jocose sallow caterpillars feed on various pine trees. Like most moths, jocose sallow moths are nocturnal. Although flying at night avoids daytime predators, nocturnal moths have to contend with spiders. Most spiders are nocturnal, and many species construct silk webs to catch night-flying insects. Although avoiding the myriad of spiderwebs in dark woods is difficult for a flying insect, moths are usually not entrapped by the webs because their wings are covered with loosely attached overlapping scales that attach to spiderwebs and break off, freeing the moth from the web. With this tricky way to avoid spiders, nocturnal moths are very successful. In North America, there are 15 times more species of moths than butterflies.

Scales on the wing of a jocose swallow

HEBREW MOTH

Latin name: *Polygrammate hebraeicum*

Family: Noctuidae

Identification: 0.9- to 1.5-inch wingspan, white forewings with black markings

Distribution: Eastern United States, west to Texas

Comments: The Hebrew moth is a member of the Noctuidae family, which are commonly known as owlet moths. Hebrew moths get their name from their forewings, which are white with black markings that are reminiscent of letters in Hebrew script. Their hindwings are dark brown or gray with an ivory-colored fringe. The abdomen of the Hebrew moth is white with black dots and lines that form the shape of a pendant necklace. Caterpillars are lavender with raised yellow bumps. They usually feed on the leaves of the black gum tree (*Nyssa sylvatica*).

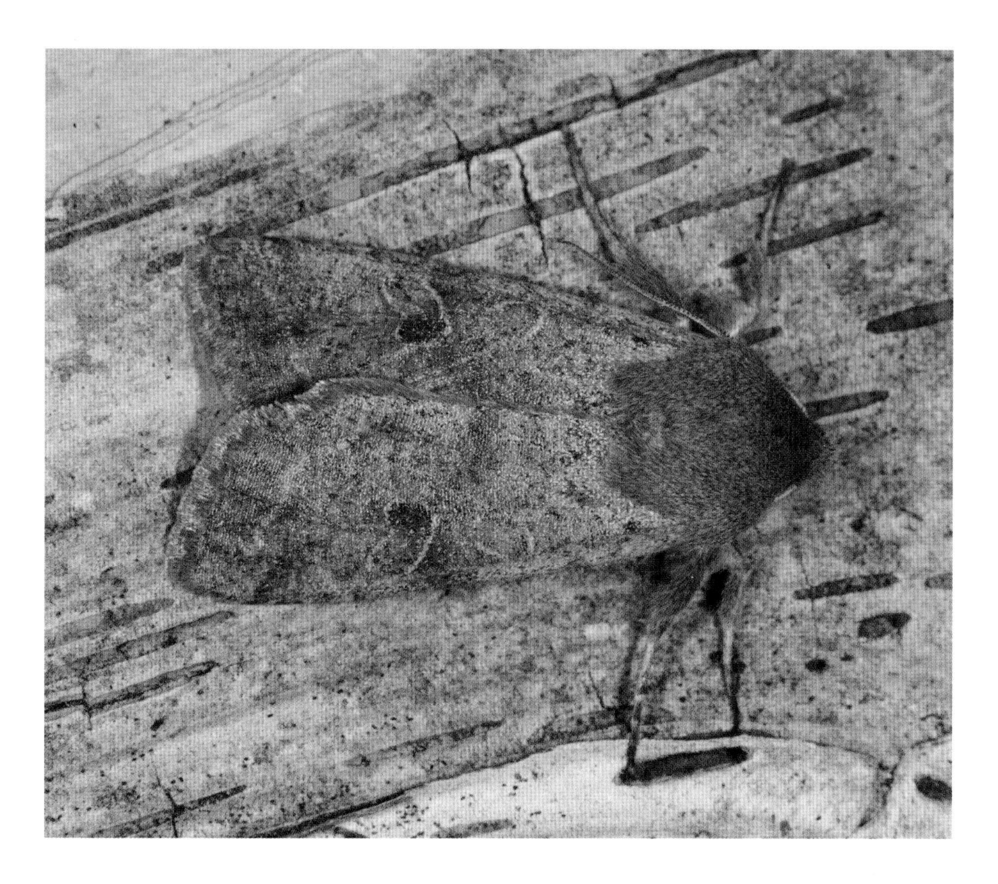

ROADSIDE SALLOW

Latin name: *Metaxaglaea viatica*

Family: Noctuidae

Identification: 2-inch wingspan, dull chestnut brown

Distribution: Eastern United States, south to Texas and Oklahoma

Comments: Roadside sallow caterpillars are pests of a number of fruit trees, especially apple, crab apple, and cherry. Caterpillars of many species of noctuid moths are known as cutworms or army worms because they often cause major damage to gardens and crops. One imported species, the cotton bollworm (*Helicoverpa armigera*), is not only a pest of cotton, but it also feeds on tomatoes, chickpeas, rice, sorghum, and cowpeas. Another noctuid moth species, the variegated cutworm (*Peridroma saucia*), is one of the most serious pests of gardens. Like the cotton bollworm, it is an imported species and also a major pest in Eurasia and Africa.

PRIMROSE MOTH

Latin name: *Schinia florida*

Family: Noctuidae

Identification: 1- to 1.2-inch wingspan, pink forewings with yellow terminal bands, yellow thorax and abdomen, white hindwings

Distribution: Alberta and Nova Scotia, south to northern Florida, west to Texas

Comments: The primrose moth appears when the flowers of primrose and gaura (a native plant with pink or white flowers) appear. Both adults and larvae are dependent on the flowers of these plants. The nocturnal pink moths are rather conspicuous, so they need a place to hide in the daytime. Primroses come in many colors including pink and yellow, which are the colors of the primrose moth, and the flowers of gaura plants are pink or white. When dawn arrives, the pink and yellow primrose moth enters a pink or red flower head first so its wings look like flower petals. The moth hides in the flower until sunset, when it is time to fly. Females lay their eggs in the flower buds of primrose and gaura plants. When the eggs hatch, the caterpillars feed inside the buds, where they cannot be seen by predators.

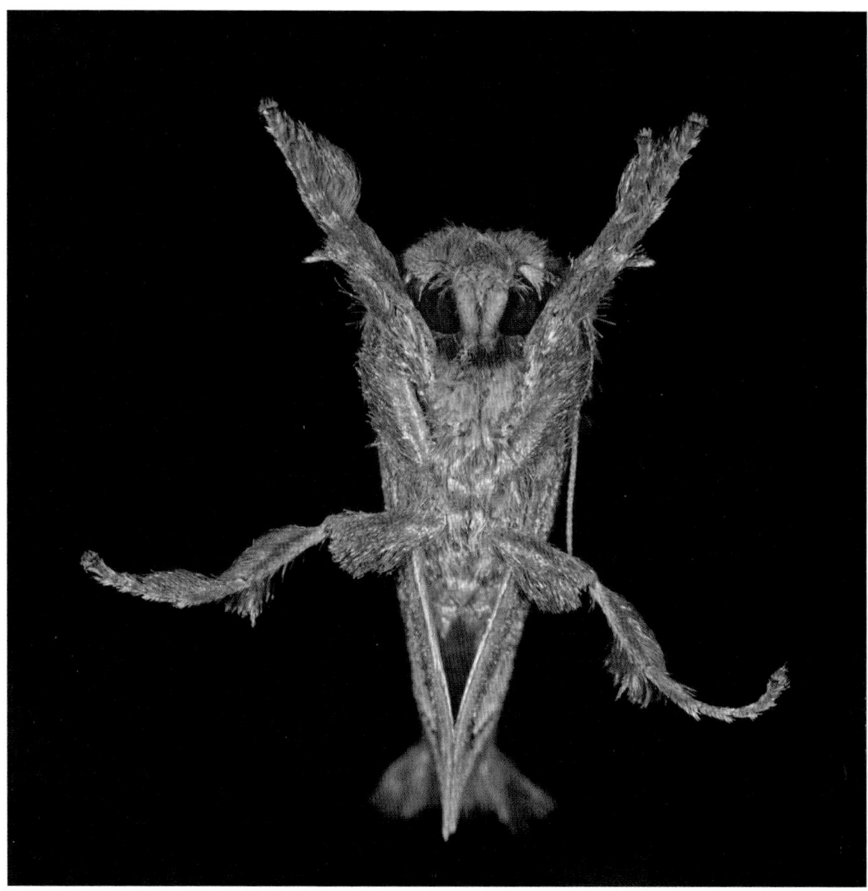

Spiny oak slug moth looking through the window

SPINY OAK SLUG MOTH

Latin name: *Euclea delphinii*

Family: Limacodidae

Identification: 0.4- to 0.6-inch wingspan, chubby, light brown with pea-green markings on forewings

Distribution: Eastern United States and southeastern Canada

Comments: The larvae of slug moths look more like slugs than caterpillars. Each ommatidium of the eyes of most insects is lined with pigment that keeps the light from passing to neighboring ommatidia. This enables insects to perceive a relatively sharp image. In nocturnal moths, the pigment moves at night such that there is no pigment between ommatidia. As a consequence, nocturnal moths see less detail at night, but their night vision is improved because light that enters one ommatidium scatters to adjacent ommatidia.

Looking at a spiny oak slug moth looking through the window

For nocturnal insects, the ability to see on dark nights is usually more important than being able to discern details.

SUZUK'S PROMALACTIS MOTH

Latin name: *Promalactis suzukiella*

Family: Oecophoridae

Identification: 0.4- to 0.6-inch wingspan, orange with 2 white stripes, prominent hairs in wingtips, prominent orange palpi, black and white legs

Distribution: Massachusetts, south to Georgia, west to Kentucky and Alabama

Comments: This beautiful little moth is native to Japan, Korea, and Taiwan. The caterpillars feed under the bark of rotting logs of chokecherry, peach, and oak trees. It is not known how this moth managed to make it across the ocean to North America, but it is hardly surprising. Suzuk's promalactus moths and caterpillars are small, and the eggs are so tiny that they go unnoticed. With thousands of container ships carrying all kinds of goods, a few eggs on some wood pallets or crates would never be noticed by customs inspectors. Fortunately, the Suzuk's promalactus moth does not seem to cause any problems.

Viewed from the front, plume moths are reminiscent of an airplane.

PLUME MOTHS

Latin name: *Pterophoridae* sp.

Family: Pterophoridae

Identification: 0.25- to 1.5-inch wingspan, featherlike wings held flat, forewings usually divided into 2 pairs and hindwings usually divided into 3 pairs

Distribution: Throughout North America

Comments: Plume moths have feathery wings. When viewed from above, the thin wings and abdomen form a T shape. When viewed head-on, these little moths have the shape of an airplane. Because of their feathery wings, predators may mistake a plume moth for dried grass and pass it by. A number of plume moths are pests. The geranium plume moth (*Platyptilia pica*) and snapgarden plum moth (*Stenoptilodees antirrhina*) damage ornamental garden plants. The artichoke plum moths *Platyptilia carduidactyla* and *Cynara cardunculus* damage artichoke crops in California.

Wings of plume moths resemble feathers.

However, not all plume moths are pests. The larvae of some species feed on invasive plants and have been used as biological controls against mistflower, horehound, groundsel, and other invasive plants.

ARCHED HOOKTIP, MASKED BIRCH CATERPILLAR

Latin name: *Deprana arcuata*

Family: Drepanidae

Identification: 0.9- to 1.6-inch wingspan, distinctive hooks on tips of forewings

Distribution: Throughout the United States, including Alaska, and most of southern Canada

Comments: Moths in the family Drepanidae are referred to as hooktips because their forewings have a prominent rear-facing hooked tip. They usually live in hardwood forests, especially in elevated areas. There are one or two generations a year. These nocturnal moths are often attracted to lights. The caterpillars of arched hooktip moths are green, brown, or purple. The four segments next to their head have bumps, each with a single black hair. They feed on white birch and alder trees, where they make communal shelters by folding leaves and fastening them together with silk. The caterpillars can produce sounds to attract other arched hooktip caterpillars or to frighten predators by shaking their bodies and rubbing or scraping their mouthparts against a leaf, or by dragging specialized anal organs on leaves.

25
FLIES

Approximately 17,000 species of flies have been described in North America. Flies live in just about every environment from forests, meadows, ponds, and backyards, to salt lakes and hot springs, to the nests of bees and wasps. Flies can best be distinguished from other insects by their having only forewings, although a few species are wingless. The hindwings of flies are modified into knob-like structures called halters that act like gyroscopes to help them navigate and perform acrobatic maneuvers. As many flies mimic bees and wasps, the best way to determine if an insect is a fly is to see if it has one or two pairs of wings. In addition, the flies that mimic bees and wasps have prominent eyes and short antennae, whereas bees and wasps have less-prominent eyes and relatively long antennae. While some kinds of bees and wasps can hover, most hovering insects are flies.

Flies develop by complete metamorphosis. Some flies, such as mosquitoes, transmit diseases such as malaria, yellow fever, and dengue fever, while others attack grain crops, orchards, and livestock. On the other hand, flies are excellent pollinators, second only to bees, and are an important food source for many animals.

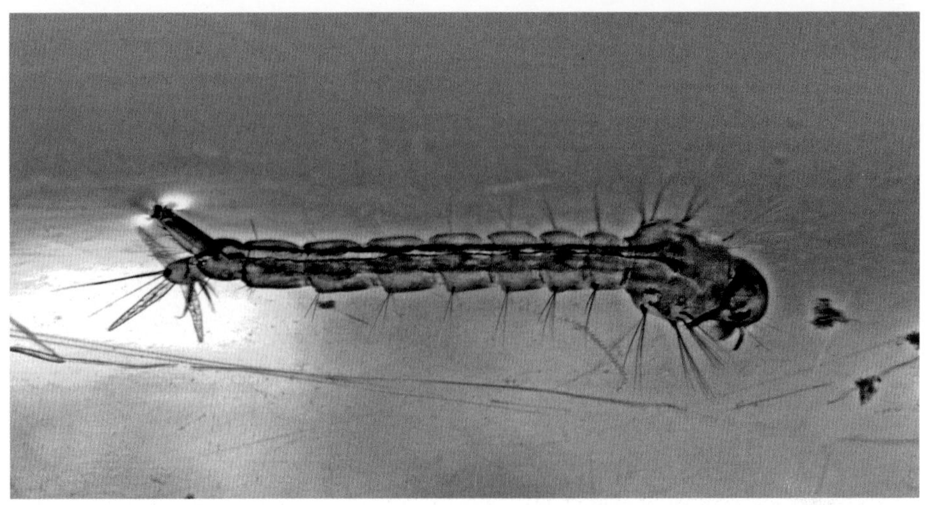

Male culex. Note anal breathing tube.

MOSQUITO LARVAE AND PUPAE

Latin name: *Culex* sp.

Family: Culicidae

Identification: Larvae have long, clear, distinct segments and breathing tube at rear; motile pupae have breathing tubes that resemble ears

Distribution: Throughout North America, but absent from coldest regions of Canada and Alaska

Comments: The aquatic larvae and pupae of *Culex* species obtain air through breathing tubes that protrude through the surface of the water. Pupae float just under the water surface because they are buoyant. Mosquito pupae are unusual because they are active. The two structures that resemble ears on the thorax of these pupae are breathing tubes that protrude through the surface of the water. They are connected to the pupa's tracheal system. The larvae and pupae of culex and other mosquito species that feed on human blood and transmit human pathogens live in standing water. They can thrive in a very small volume of water, such as a tin can or small puddle. I have read that they can even mature in the water that is in a bottle cap. Because they feed on human blood, they are most common where people live. It is difficult to eliminate mosquitoes because they can survive in such a small volume of water.

Culex pupa

Female culex

MOSQUITOES

Latin name: *Culex* sp.

Family: Culicidae

Identification: 0.2 to 0.4 inch long, clear wings

Distribution: Throughout North America, but absent from coldest regions of Canada and Alaska

Comments: The ability to detect heat and carbon dioxide enables female mosquitoes to find the animals on which they feed. Culex mosquitoes are the principal vectors of West Nile virus in North America. They can acquire the virus by feeding on the blood of birds or horses that harbor it. The buzzing sound that female mosquitoes produce is not made to annoy us when we are trying to get to sleep and certainly not to wake us up so that we can swat them. Rather, the buzzing is made to attract a mate. The prominent antennae of male mosquitoes are covered with numerous setae. When the male mosquito detects the buzzing of a female of his species, he orientates his body so that the intensity of the signal is the same on both of his antennae. In this way, he is able to fly in the direction of the female even when she changes her position.

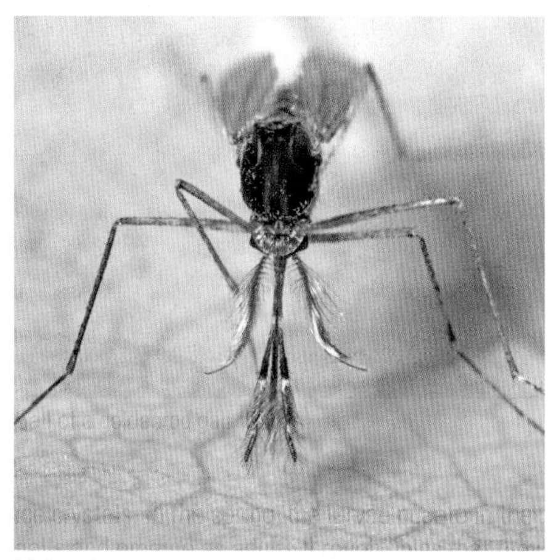

Male culex

TIGER MOSQUITO, ASIAN TIGER MOSQUITO, FOREST MOSQUITO

Latin name: *Aedes albopictus*

Family: Culicidae

Identification: 0.2 to 0.4 inch long, black-and-white-striped legs and body

Distribution: Southern New England, south to Florida, west to Texas and Illinois

Comments: Tiger mosquito larvae were accidentally imported to North America from Asia in 1997 in rainwater that had accumulated in old tires. Adults are usually active in the daytime, and their bite can be painful. Males and females feed on nectar and, being mosquitoes, females also feed on human blood. However, unlike most other mosquito pests, they also feed on the blood of other mammals as well as birds. Because they feed on the blood of mammals other than man, they can acquire pathogens, such as West Nile virus, which can infect both man and other mammals. The good news is that because female tiger mosquitoes feed on so many species of mammals and birds, they are less likely to bite people. In Asia, this species of mosquito transmits yellow fever, dengue fever, Zika, and chikungunya fever as well as some pathogenic nematodes, such as roundworms and threadworms. Fortunately, none of these diseases are common in North America.

FUNGUS GNATS, DARK-WINGED FUNGUS GNATS

Latin name: *Sciara* sp.

Family: Sciaridae

Identification: 0.2 to 0.3 inch long, dainty, dark body and wings

Distribution: Throughout North America

Comments: If you see some of these flies on a houseplant, you might not recognize them as flies because they are so small and dainty, and they fly so weakly. They are often seen on houseplants or in greenhouses flitting from plant to plant. Fungus gnats are often considered pests on houseplants. Although they do not damage healthy plants, when they are present, it may indicate that the plants are overwatered, as fungus gnats feed on rotting roots and fungi that grow in wet soil. They can usually be eliminated by covering the soil in flowerpots with a layer of sand.

Crane fly hanging on a holly bush

CRANE FLIES

Latin name: *Tipulidae* sp.

Family: Tipulidae

Identification: 0.6 to 0.9 inch long, brown, long legs, look like giant mosquitoes

Distribution: Throughout North America

Comments: No! That isn't a giant mosquito. It's a crane fly. There are no giant mosquitoes—they are all about the same size. Crane flies do not bite and there is no reason to fear them. Many species of crane flies look so much alike that only an expert can identify them. They are often confused with the hangingflies of the Bittacidae family, which are not actually flies, because they hang on twigs like hangingflies and both crane flies and hangingflies have long, thin legs. Although they are similar in appearance, crane flies have one pair of wings and hangingflies have two. Hangingflies also have spurs on their legs, while spurs are absent from the legs of crane flies. Although they start out with six legs, the long delicate legs of crane flies often break off. However, losing a leg or two doesn't seem to bother them. Larvae of most species of crane flies live in wet soil, where they feed on fungi and decaying plants.

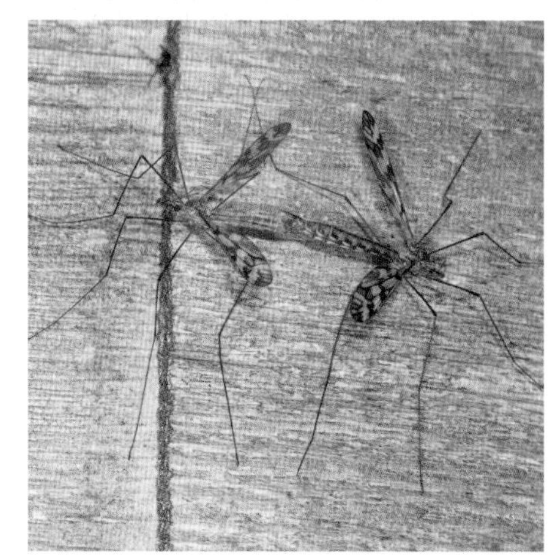
Crane flies mating. Female on right

WINTER CRANE FLIES

Latin name: *Trichocera* sp.

Family: Trichoceridae

Identification: 0.2 to 0.4 inch long, long legs, resemble crane flies

Distribution: Throughout North America

Comments: These dainty long-legged flies resemble crane flies but are smaller than most crane flies and a bit sturdier. They usually have all their legs, whereas crane flies are often missing a leg or two. Unlike crane flies, winter crane flies have ocelli, although they are difficult to see without a magnifying glass because these flies are small and have small heads. The winter crane fly flies not only in the spring, summer, and fall but also sometimes in the winter, which makes them one of the few insects that are up and about in the wintertime, hence their name. The larvae of winter crane flies live in decaying plant matter and are most common in the colder months.

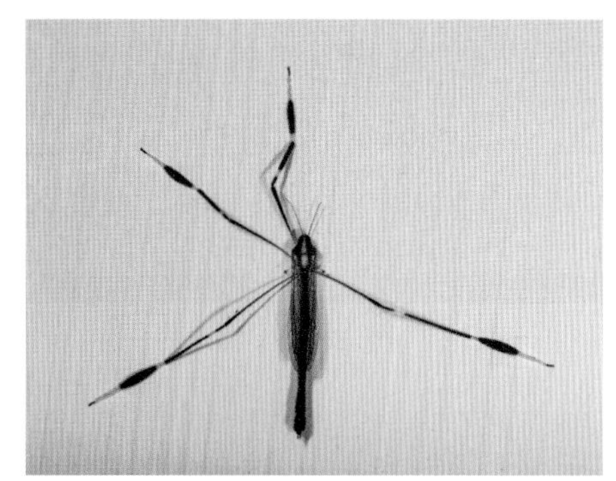

Phantom fly, missing two legs

PHANTOM CRANE FLIES

Latin name: *Bittacomorpha* sp.

Family: Ptychopteridae

Identification: 0.8 to 1.2 inches long, swollen black legs with white markings, resemble crane flies

Distribution: Eastern United States, west to Arizona; Newfoundland, west to Manitoba

Comments: Phantom crane flies have a ghostly appearance as they float in the breeze, lifted by their long, flattened legs. Their larvae live in mud along the shores of ponds and swamps, where they feed on decaying organic matter. They breathe through a long respiratory tube on their posterior that they extend above the surface of the mud.

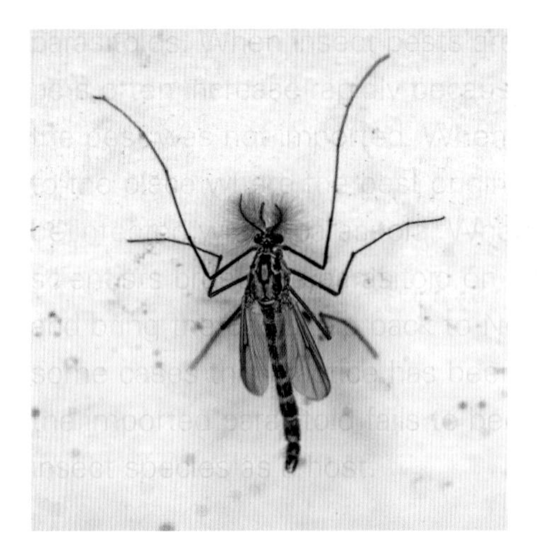

MIDGES, GNATS

Latin name: *Chironomus* sp.

Family: Chironomidae

Identification: 0.06 to 0.3 inch long; males have plumose antennae; resemble mosquitoes, but at rest, they hold their wings to the side, whereas mosquitoes fold their wings over their back

Distribution: Throughout North America

Comments: Midges are some of the most common insects. They are often found in large numbers near water, where their larvae live. I used to be a runner, but I never ran near the lake that was close to our house, because if I did, I would constantly be inhaling these dainty little flies. Although they resemble mosquitoes, they do not bite. They can be distinguished from mosquitoes because they hold their wings out to the side rather than folding them over their back, and because they hold their forelegs out straight in front of their body in a characteristic manner. The aquatic larvae of most midges contain hemoglobin in their hemolymph, which is most unusual, but these larvae need an oxygen-carrying pigment because they live in mud deep at the bottom of lakes, where there is very little oxygen. *Chironomus* larvae are often called bloodworms because of their red color imparted by the hemoglobin.

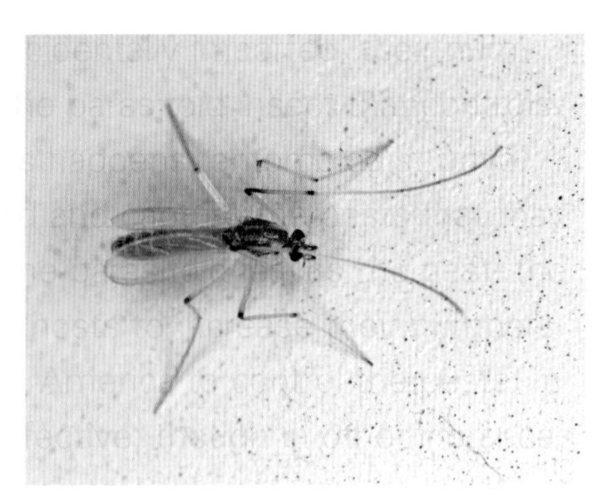

GREEN MIDGES

Latin name: *Tanytarsus* sp.

Family: Chironomidae

Identification: 0.06 to 0.3 inch long, males have plumose antennae, at rest they hold wings to the side, resemble mosquitoes

Distribution: Throughout North America

Comments: Although most midge larvae build silk tubes and live in the mud at the bottom of streams, rivers, or lakes, green midge larvae are free-living and propel themselves by wiggling. As these larvae are an important food source for fish, anglers often use "flies" that resemble midge larvae to catch trout. To determine if a stream, river, or lake is polluted, scientists look for midge larvae, as well as larvae of mayflies and stoneflies. Contaminated water contains very few of these larvae, while uncontaminated water contains many. Counts of midge larvae on successive years can determine whether a stream, river, or lake is becoming polluted.

A swarm of biting midges

BITING MIDGES, PUNKIES, NO-SEE-UMS

Latin name: *Ceratopogonidae* sp.

Family: Ceratopogonidae

Identification: 0.03 to 0.15 inch long, brown, often in swarms

Distribution: Aquatic, semiaquatic, and mountainous habitats across North America

Comments: No-see-um is an appropriate name for these tiny flies because they are so small you may not see them. In fact, they are so small that they can get through most screens. The females of many species feed on the blood of animals. Although some feed on human blood, many kinds of biting midges feed on the blood of other animals. A bite from these little flies can cause redness and itching, as the saliva causes a local allergic reaction. Male punkies form swarms of up to thousands of individuals. They usually swarm in the same location each year, using landmarks such as boulders or trees to identify their gathering place. Females use the same landmarks and fly into the swarm. They quickly find a male, and the couple falls to the ground as they mate. After mating, the females fly off to lay their eggs, and the males rejoin the swarm, which is why the swarms of these and other insects are composed primarily of males.

Male biting midge

WINDOW GNAT, WOOD GNAT

Latin name: *Sylvicola fenestralis*

Family: Anisopodidae

Identification: 0.3 to 0.4 inch long, long legs, wings with black pattern, resembles a crane fly, females larger than males

Distribution: Throughout North America

Comments: The name of this species of gnat, *fenestralis,* derives from the Latin word *fenestra,* which means "window." In fact, these little flies often enter homes and fly to windows, hoping to get out of the stuffy building. They feed on decaying plants and fungi, so they won't eat anything in your home, and they don't bite. I suggest that you leave them alone or, even better, admire them.

Window gnats mating. Female on left

MOTH FLIES, DRAIN FLIES

Latin name: *Clogmia* sp.

Family: Psychodidae

Identification: 0.08 to 0.15 inch long, mothlike, stocky, hairy wings and body

Distribution: Throughout North America

Comments: The body and wings of moth flies are hairy, which makes them look a bit like moths, thus the name moth flies. Although they live near ponds and swampy areas, you are most likely to encounter them in kitchens and bathrooms. Because they are small, they can enter through cracks and crevices, and they love bathroom sinks and especially urinals in men's rooms that are relatively open to the outside. Sometimes moth flies congregate in large numbers in sewage treatment plants. Females deposit their eggs in drains, where the larvae feed on gunk in the drain trap, which is why you can find adult flies in bathrooms and kitchens. Although they don't bite, some people find them a nuisance, especially if there are many of them. The best way to get rid of moth flies is to keep your bathroom and kitchen drains clean so that gunk does not accumulate. There are several commercial products containing bacteria or enzymes that break down gunk in drains.

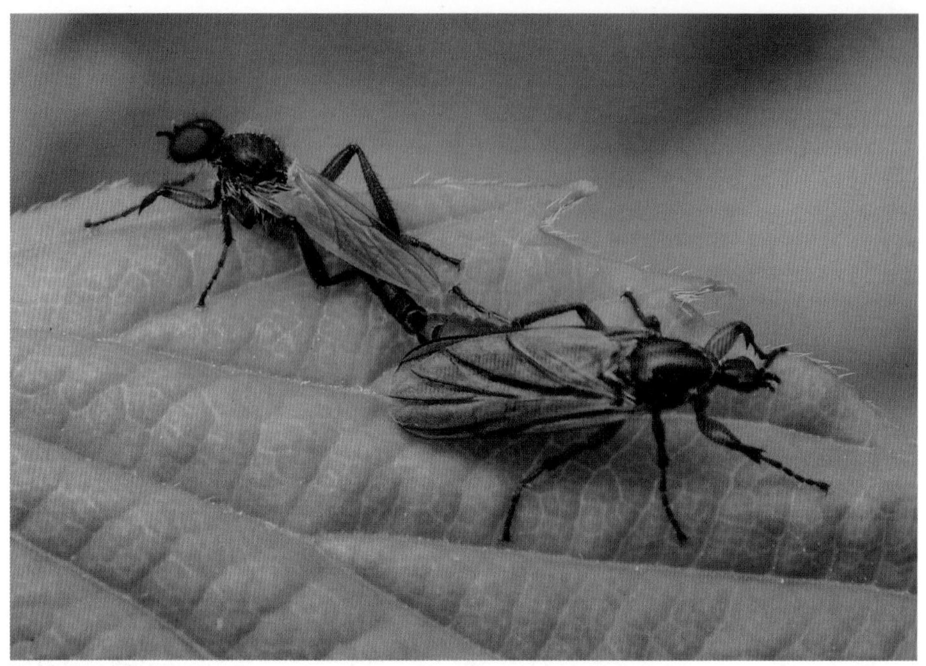

March flies mating. Female below

MARCH FLIES

Latin name: *Bibio* sp.

Family: Bibionidae

Identification: 0.3 to 0.5 inch long, blackish, males have a large head and large eyes, larger females have a small head and small eyes

Distribution: Throughout the United States and southern Canada

Comments: March flies of the genus *Bibio* are very common in the spring and early summer throughout North America, where they fly low over fields and can sometimes be seen in large swarms. Adult flies do not feed. Larvae live in the soil, where they scavenge plant roots and dead vegetation. Pairs of flies often remain attached for many hours after mating. Insemination does not take that long, but holding on to your mate is a good way for a male to ensure that the female will not fly off and mate with another male. Males are equipped with forcepslike structures called claspers that enable them to hold on to their mate even if she tries to walk or fly away. Flies of a species that occurs in the South, *Plecia nearctica*, are called love bugs because they are so often found in pairs. Sometimes they are so numerous that they can make quite a mess on the windshields of cars.

Male march fly on screen

Green bottle flies. Larger female on right

GREEN BOTTLE FLY

Latin name: *Lucilia sericata, Phaenicia sericata*

Family: Calliphoridae

Identification: 0.2 to 0.4 inch long, iridescent green or greenish gold

Distribution: Across North America and the world, wherever people live

Comments: Green bottle flies are a common type of blowfly that live less than a month as adults, during which time a female may lay several thousand eggs. Depending on the temperature, humidity, and availability of food, in a few weeks, the eggs usually hatch mature and become adults. Many kinds of flies transmit human and animal diseases by biting. However, flies like the green bottle fly can also transmit bacteria and viruses because of the way they feed. Houseflies, bottle flies, some species of flesh flies, and certain other kinds of flies do not eat solid food. Instead, they regurgitate saliva containing digestive enzymes that liquefy solid food. They then lap up the liquid. When they fly from one food source to another, these flies may transfer pathogens that they acquired on their bodies or in their saliva at a garbage dump or from a dead animal to food at an open market, picnic table, or kitchen.

DEERFLIES

Latin name: *Chrysops* sp.

Family: Tabanidae

Identification: 0.3 to 0.5 inch long, yellow with black markings, colorful eyes

Distribution: New Brunswick to Florida, west to British Columbia and Arizona

Comments: Male deerflies feed on nectar. Females feed on blood and require a blood meal and a few days to digest the meal that nourishes their eggs before they can be laid. Females use their sharp cutting mouthparts to wound animals, including humans. They then lap up the blood. With their keen eyesight and robust flight, when they are feeding, these flies can detect the slightest movement and be airborne in an instant. However, even when they detect danger, hungry female deerflies often continue feeding because most animals have no hands with which to swat at them. However, with humans, a decision not to take off can be fatal. Deerfly bites can be painful and may cause an allergic reaction to the salivary secretions that are released when the fly feeds, and secondary infections can occur if the bites are scratched. Although deerflies usually do not transmit diseases in North America, they can transmit tularemia and anthrax in some developing countries.

Deerfly eye

Striped horsefly, *Tabanus lineola*

HORSEFLIES, GADFLIES

Latin name: *Tabanus* sp.

Family: Tabanidae

Identification: 0.7 to 1.4 inches long, brightly colored eyes

Distribution: Throughout North America

Comments: Depending on their species, female horseflies lay from 30 to as many as 1,000 eggs in the mud along ponds, wetlands, and streams. When the eggs hatch, the legless larvae feed on decaying organic matter and small invertebrates in the mud. They remain as larvae for 1 to 3 years, depending on the species, before they crawl to drier land and pupate. Adults are particularly agile and fast on the wing. Both males and females feed on nectar, but females also feed on blood by wounding animals with their knifelike mouthparts and lapping up the blood. Horseflies are a serious pest to livestock because when they are in large numbers, they can consume enough blood from cattle and horses to render them weak and reduce milk production in cows. Horseflies also bite humans, which can be very painful and result in an allergic reaction to the fly's saliva. Although they do not transmit diseases in North America, horseflies are vectors of a number of human and animal diseases in tropical and semitropical regions of the world.

Knifelike mouthparts of a female horsefly

Female greenhead

GREENHEAD

Latin name: *Tabanus americanus*

Family: Tabanidae

Identification: 0.5 to 0.7 inch long, robust, bright green eyes

Distribution: Coastal regions from eastern Canada to Florida and the Gulf Coast west to Texas

Comments: Greenheads are a species of horseflies that breed in salt marshes. Males feed on nectar, but females feed on blood and give a nasty bite with their knifelike mouthparts. There used to be so many greenheads on the coast of New England that early settlers often had to work at night to avoid being bitten. In some places, greenheads were so prevalent that they could cause farm animals to become anemic due to loss of blood from greenhead bites. The bite was so painful that oxen and horses sometimes ran off after being bitten and injured themselves. Before the widespread use of traps, greenheads also affected the tourist trade on the New England coast. These simple traps consist of a dark blue box with a funnel-like entry on the bottom. Sometimes bovine blood or a synthetic attractant is used as bait. Greenheads fly up into the trap, where they cannot escape and eventually die.

Eyes of a female greenhead

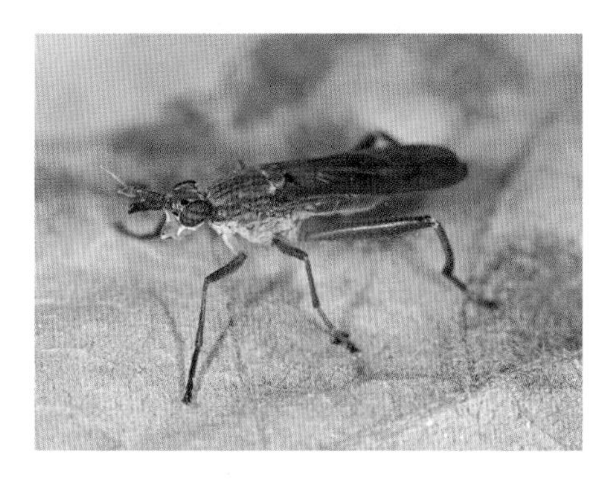

MARSH FLIES, SNAIL-KILLING FLIES

Latin name: *Tetanocera* sp.

Family: Sciomyzidae

Identification: 0.3 to 0.5 inch long, slender, brown, concave face, long antennae

Distribution: East and West Coasts

Comments: Marsh flies feed on nectar. They can often be observed running along the shores of streams, rivers, and ponds. Eggs of some species of marsh flies are scattered on the shore, whereas other species lay their eggs on land snails or slugs. When the larvae of eggs that are laid on land hatch, they enter the water and seek out the freshwater snails that they feed on. After consuming the snail, they pupate in the snail's shell.

FOUR-BARRED KNAPWEED GALL FLY, SEEDHEAD FLY

Latin name: *Urophora quadrifasciata*

Family: Tephritidae

Identification: 0.07 to 0.1 inch long, 2 prominent black V-shaped markings on each wing, white spot on thorax

Distribution: Throughout the United States and southern Canada

Comments: Since the four-barred knapweed gall fly was introduced into the United States, they have spread throughout most of North America. Knapweeds are invasive weeds that establish quickly in overgrazed land as well as grasslands, where they outcompete native vegetation. Female flies use their retractable ovipositor to penetrate flower buds, where they lay their eggs. When the larvae hatch, they induce the flower to form a gall where they mature, pupate, and emerge as adults. This destroys the flower so that the plant cannot reproduce. Four-barred knapweed gall flies have two generations per year.

FRUIT FLIES

Latin name: *Rhagoletis* sp.

Family: Tephritidae

Identification: 0.15 to 0.2 inch long, black with black bands on their clear wings

Distribution: Eastern United States and southern Canada

Comments: There are twenty-one species of the genus *Rhagoletis* in North America. They all have dark bands on their wings, which look like the front legs of the jumping spiders that they mimic. Male flies wave their wings as they court females, which is why flies of the genus *Rhagoletis* are sometimes called peacock flies. Several species are pests of various fruits. The cherry fruit fly (*Rhagoletis cingulata*) is a serious pest of cherries. It also has been imported to Europe. Before apples were imported to North America, *Rhagoletis pomonella* fed on hawthorn, but now it is a serious pest in apple orchards. The walnut husk fly (*Rhagoletis completa*) bores into the shell of walnuts, making the walnut deformed with rotting husks.

BEE FLIES

Latin name: *Systoechus* sp.

Family: Bombyliidae

Identification: 0.2 to 0.4 inch long, swept-back wings, exceptionally long proboscis

Distribution: Throughout North America

Comments: Adult bee flies feed on nectar and pollen and are important pollinators. Flies of the genus *Systoechus* are especially pubescent and may easily be mistaken for little bees. They have an exceptionally long proboscis that they hold straight because they cannot retract or curl it up as butterflies do. They are strong fliers and characteristically hover over flowers as they feed and alternatively rest in the sun. The larvae of *Systoechus* sp. are parasitoids. However, unlike the larvae of most parasitoids that are limited in the species of insects that they feed on, bee flies of the genus *Systoechus* can have many different hosts. Females of some species of *Systoechus* lay their eggs near the nests of solitary bees or wasps. Females of other species scatter their eggs, leaving their larvae to find the burrows of the bees, wasps, beetles, and other insects on which the parasitoid larvae feed.

BLACK-TAILED BEE FLY

Latin name: *Bombylius major*

Family: Bombyliidae

Identification: 0.5 to 0.7 inch long, hairy black and yellow, long proboscis, front of wings black, rear of wings transparent

Distribution: Throughout North America

Comments: These bee flies are convincing bee mimics. They are effective pollinators, especially at high altitudes where many other pollinating insects are scarce. They look a bit menacing with their long proboscis but are harmless. The parasitoid larvae of these bee flies have many hosts, including solitary bees and wasps and some species of beetles. Females fly over nests of hosts and drop eggs into or adjacent to the nest. When the eggs hatch, the larvae crawl into the nest and feed on the food stores and host larvae. The host bee or wasp usually ignores the bee fly larvae, even though they are feeding on her young. If they can't find a bee nest, black-tailed bee flies will lay their eggs on a flower that the host might visit. When the eggs hatch, the larvae wait for a bee or wasp and grab on to her for a ride back to her nest.

TIGER BEE FLY

Latin name: *Xenox tigrinus*

Family: Bombyliidae

Identification: 0.3 to 0.6 inch long, stout with short abdomen, swept-back angled wings with distinctive black pattern

Distribution: East Coast of the United States and southwest to Arizona

Comments: Adult tiger bee flies feed on nectar, but their larvae are parasitoids of carpenter bee larvae. While the mother carpenter bee is away, tiger bee flies will check to see that the bee has laid eggs in the nest. If eggs or larvae are present, the tiger bee fly lays her own eggs next to, or sometimes in, the nest. When they hatch, the tiger bee larvae crawl into the carpenter bee's nest and gradually consume the carpenter bee larvae. When the mother carpenter bee returns with food for her young, she puts down the food, ignores the tiger bee fly larvae that are consuming her young, and flies off to get more food. The tiger bee fly larvae mature, pupate in the carpenter bee's nest, and fly off.

TACHINID FLIES, TACHINIDS

Latin name: *Leschenaultia* sp.

Family: Tachinidae

Identification: 0.3 to 0.5 inch long, large black fly, prominent long black setae, concave face

Distribution: Throughout the United States and southern Canada

Comments: Most tachinid flies are robust fliers. Adults feed on flowers, but their larvae are parasitoids of various insects. Female flies have evolved various ways to get their larvae into the host. The hosts of tachinid flies of the genus *Leschenaultia* are caterpillars. Adult females lay their eggs on leaves. If a caterpillar happens to eat some of the fly's eggs along with the leaf and does not kill the eggs as it feeds, the larvae will hatch inside the caterpillar, pupate, and eventually emerge as adults. As the larvae slowly feed on the caterpillar, they are careful not to eat any vital organs so they will not kill the caterpillar before they are ready to molt. The caterpillar eventually dies from the injuries.

TACHINID FLIES, TACHINIDS

Latin name: *Gymnosoma* sp.

Family: Tachinidae

Identification: 0.2 to 0.3 inch long, red or orange abdomen with distinctive black pattern

Distribution: Throughout the United States and Ontario

Comments: *Gymnosoma* flies are parasitoids of stink bugs and shield bugs. After mating, female flies lay their eggs on the host. When the larvae hatch, they crawl into the stink bug or shield bug through the soft areas between its abdominal segments. The larvae feed on the fat bodies of their host, which sterilizes the insect but usually does not kill it. A number of species of tachinid flies have been imported into North America to control pests. This is a somewhat risky business, not only because the tachinid fly may not become established, but also because it may parasitize insects other than the one it was imported to control. For example, the tachinid fly *Compsilura concinnata* was imported to control gypsy moths, but the program hasn't been successful because the fly is a generalist and lays its eggs on almost any species of moth, butterfly, or sawfly.

TACHINID FLIES, TACHINIDS

Latin name: *Trichopoda* sp.

Family: Tachinidae

Identification: 0.3 to 0.4 inch long, black with yellow markings, slender abdomen, brown eyes and wings, black legs with yellow feet, hind legs with long setae, large halters

Distribution: Throughout the United States and southern Ontario

Comments: These flies are first seen in late spring or summer feeding on flowers, especially meadow-sweet and Queen Anne's lace, or hovering over plants looking for the adult or nymph squash bug, leaf-footed bug, stink bug, or another hemipteran on which to lay their eggs. They lay several eggs on the back of the bug's thorax or abdomen, where it is unable to remove the eggs. When the larvae hatch, they burrow into their host and feed on the tissues of the bug. If there are several larvae in a bug, only one survives. When it is ready to pupate, the larva crawls from the bug and pupates in the ground. Soon after the larva leaves, the host dies. There are up to three generations a year. Second instar larvae overwinter inside the body of their overwintering host.

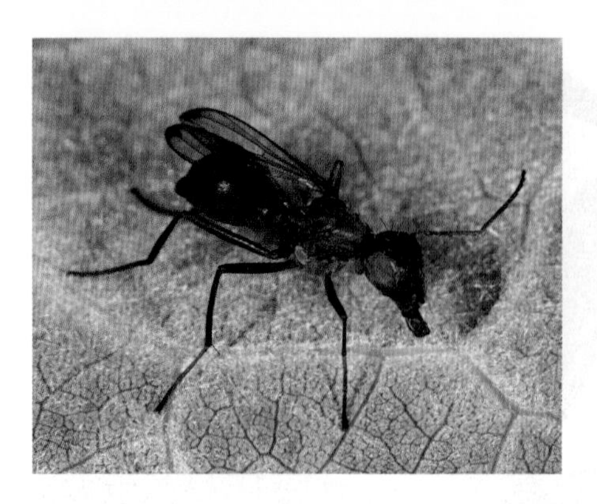

STILT-LEGGED FLIES

Latin name: *Taeniaptera* sp.

Family: Micropezidae

Identification: 0.3 to 0.5 inch long, black, long thin legs, thin waist, ant mimic

Distribution: Throughout the United States and southern Canada

Comments: Stilt-legged flies are ant and wasp mimics. Flies of the genus *Taeniaptera* mimic ants because birds usually do not eat ants, as they are sour-tasting. Like many ants, they are black and have a thin waist and round abdomen. When they walk or stand, they hold their front legs up so they look like ant antennae. Adults and larvae feed on dung, decaying vegetation, and fungi.

TRANSVERSE FLOWER FLY

Latin name: *Eristalis transversa*

Family: Syrphidae

Identification: 0.3 to 0.4 inch long, black with yellow markings on abdomen, resembles a bee

Distribution: Eastern United States and southern Ontario, other species of *Eristalis* across North America

Comments: These little bee-mimicking flies are widespread across eastern North America. Not only do they resemble bees, but they hover like bees, and their flight patterns are similar to that of bees. Males can be distinguished from females by their larger eyes that almost touch, whereas the eyes of females are separated. The larvae of flies of the genus *Eristalis* are called rat-tailed maggots because the siphon that the larvae breathe through, which is usually several times the length of the maggot, resembles the tail of a rat. Rat-tailed larvae of flies of the genus *Eristalis* live in polluted water or dead mammals. When they are ready to pupate, they crawl out of the water or dead animal and pupate in wet soil.

FLOWER FLIES, HOVER FLIES, SYRPHID FLIES, DRONE FLIES

Latin name: *Platycheirus* sp.

Family: Syrphidae

Identification: 0.4 to 0.5 inch long, shiny black or orange and black with large red eyes

Distribution: Throughout North America

Comments: Flower flies are also called hover flies because they hover over flowers as they feed on nectar and pollen. *Platycheirus* larvae are legless maggots and, like the larvae of certain other hover fly species, feed on aphids. You might wonder how these slow-moving legless maggots find aphids. The answer is their mother finds the aphids for them. Typically thousands of aphids feed on the same plant. After mating, female *Platycheirus* fly from plant to plant, hovering over leaves and using their keen eyesight to find groups of aphids. When they find a leaf that is covered with aphids, they hover over the leaf as they lay their eggs. When the eggs hatch, the larvae wander around the leaf in the night, devouring aphids.

FLOWER FLIES, HOVER FLIES, SYRPHID FLIES, DRONE FLIES

Latin name: *Allograpta* sp.

Family: Syrphidae

Identification: 0.4 to 0.5 inch long

Distribution: Throughout North America

Comments: Flies of the genus *Allograpta* are bee mimics. Like other genera of flower flies, they feed on nectar and pollen—nectar for their energy source and pollen as their source of protein. Flower flies are important pollinators of food crops. Because they have short mouthparts, they tend to feed on flowers that are more open, as nectar and pollen can be easily reached. Different species of hover flies have different preferences for flowers, but many prefer white and yellow flowers. Smell and other nonvisual clues appear to be important in flower selection. Although the larvae of most species of *Allograpta* feed on soft-bodied insects, the larvae of some species are leaf miners, stem borers, or feed on pollen.

NARCISSUS BULB FLY

Latin name: *Merodon equestris*

Family: Syrphidae

Identification: 0.5 inch long, heavy-bodied bee mimic, black and yellow, hairy abdomen

Distribution: Throughout the United States and southern Canada

Comments: Since *Merodon equestris* was imported from Europe many years ago, these flies have spread throughout most of North America. Most bee flies do not cause any problems, but *Merodon equestris* is an exception. In the spring female flies dig a hole in the earth next to a narcissus (daffodil or jonquil) bulb and lay an egg on the stem of the bulb. Females lay one egg on each of a hundred or so bulbs. When the larvae hatch, they feed on the bulb all fall and winter. They crawl from the bulb in the spring, pupate in the ground, and emerge as adult flies in April or May. The bulb is often killed, or the plant does not flower.

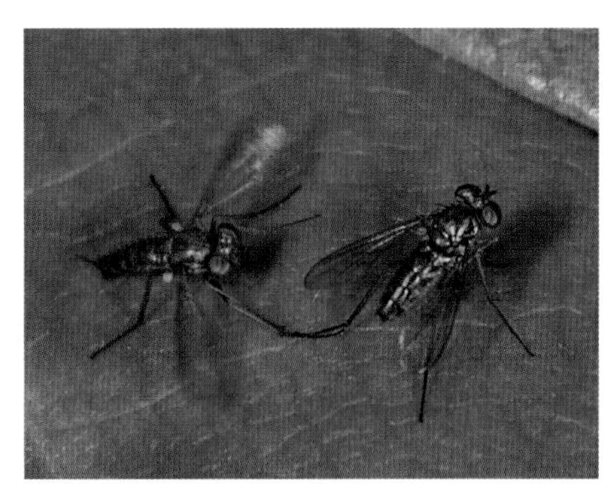

Male long-legged fly (on left), courting a female

LONG-LEGGED FLIES

Latin name: *Condylostylus* sp.

Family: Dolichopodidae

Identification: 0.1 to 0.2 inch long, metallic coppery green, long slender legs, large red eyes

Distribution: Eastern North America

Comments: These little metallic-colored flies are predators of small soft-bodied arthropods, especially mites, aphids, and other small flies, but also feed on nectar. Like many other flies, long-legged flies have huge eyes and most of their brain is devoted to interpreting visual images. In order for a male fly to win mates, they do a little dance. As a male dances, he flutters his wings wildly as he approaches a potential mate. If she is impressed, she will mate with him. If not, she flies off. The predatory larvae of *Condylostylus* live in soil, under bark, or in freshwater.

LONG-TAILED DANCE FLIES

Latin name: *Rhamphomyia* sp.

Family: Empididae

Identification: 0.15 to 0.25 inch long, humpbacked, small head, long slender legs, orange eyes

Distribution: Throughout the United States and southern Canada

Comments: Dance flies are so named because males form swarms in which the flies move up and down. Dance flies feed on insects, and males in the swarm often carry a dead insect that they captured and killed in order to entice a female. Females fly into the swarm and find mates and then fly off with them. Females of the genus *Rhamphomyia* are completely dependent on males for the meal that they require before their eggs can develop. In their courtship ritual, males present a "gift" of a dead insect, usually a mosquito or a march fly, to their prospective mate for her to eat while they mate. Females who have mated and laid their eggs often inflate their abdomen with air so as to appear as though they have eggs in order to obtain a free meal from a male.

PICTURE-WINGED FLIES

Latin name: *Callopistromyia* sp.

Family: Ulidiidae (formerly Otitidae)

Identification: 0.2 to 0.3 inch long, wings with black pattern, gray body with black spots

Distribution: Northeastern United States

Comments: This group of flies is called picture-winged flies because of the intricate patterns on their wings. They are usually found in moist places, where adult flies and larvae feed on decaying plant matter. Most species of picture-winged flies have no effect on humans, but two species are pests. The sugar beet root maggot (*Tetanops myopaeformis*) is a serious pest of sugar beets in eastern North Dakota and northwestern Minnesota. The other pest is the cornsilk fly (*Euxesta stigmatias*), which attacks the ears of sweet corn in Florida. Sometimes entire fields of sweet corn are so damaged by these picture-winged flies that they are not harvested.

Picture-winged fly. Note white halters.

PICTURE-WINGED FLY

Latin name: *Delphinia picta*

Family: Ulidiidae

Identification: 0.3 to 0.4 inch long; black wings with 2 transparent triangle areas, one on front and one on back

Distribution: Throughout eastern North America, west to Kansas and Minnesota

Comments: *Delphinia picta,* as well as other flies in the family Ulidiidae, are best recognized by the patterns of bands and spots on their wings. The halters of *Delphinia picta* are particularly obvious. Their larvae feed on decaying vegetation on the surface of the ground or on rotting fruit. They also feed on the fermented frass of the locust borer left on black locust trees. Unlike most insect genera that have a number of species, *picta* is the only species in the genus *Delphinia. Picta* is from the Latin word for "painted." Their picture wings are used during courtship. Females wave their wings gently, and males respond by flickering their wings. Females lay up to 500 eggs in decaying plant matter. They have one generation a year from May until June and overwinter as mature, third instar larvae.

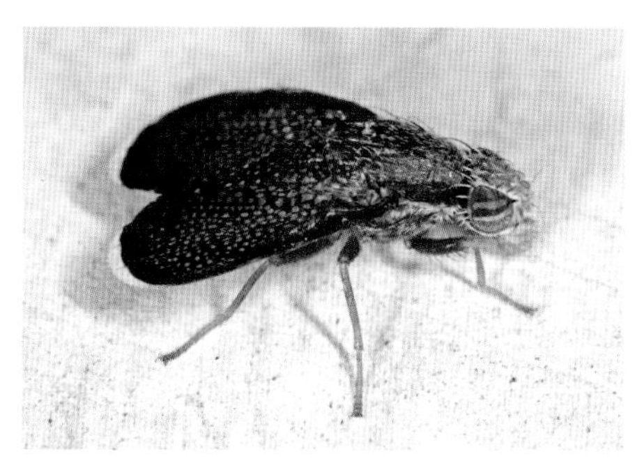

SOLDIER FLIES

Latin name: *Sargus* sp.

Family: Stratiomyidae

Identification: 0.4 to 0.6 inch long, wide head, large eyes, Y-shaped antennae

Distribution: Throughout North America

Comments: Soldier flies are sluggish medium-size flies that spend a lot of their time resting on flowers and plants. Adult flies have vestigial mouthparts and do not feed. Their larvae feed on decaying plant matter. Larvae of the black soldier fly, *Hermetia illucens*, are raised to feed poultry and pigs. Black soldier fly larvae and the compost bins that they are kept in are for sale on the Web. The wooden bins are a few feet long. They have a removable screen-covered top, a tray, and containers the size of small wastebaskets on either side. Organic waste or fresh manure is placed in the bin along with some black soldier fly larvae. The larvae feed, pupate, and become adults. The females then lay eggs that develop into more larvae. Since the larvae are active, some of them crawl to the edges of the tray and fall into the bins, which are collected regularly. The larvae are fed to farm animals, and new garbage or manure is added to the bin.

FRUIT FLIES

Latin name: *Eutreta* sp.

Family: Tephritidae

Identification: 0.2 to 0.5 inch long, black wings with clear spots and white wingtips, red eyes with yellow stripes

Distribution: Throughout North America

Comments: Some fruit flies of the genus *Eutreta* lay their eggs on plants of the Asteraceae family, while others lay their eggs on plants of the verbena family. When the larvae hatch, they induce the plants to produce galls on their stems or roots that can harm the plant. One of the most serious fruit fly pests is the Mediterranean fruit fly, medfly (*Ceratitis capitata*). This sub-Saharan fruit fly has spread all over the world, where it attacks many kinds of fruit, especially citrus. One method of control used in California is to grow millions of male medflies and irradiate them. This damages the chromosomes of the cells that will become sperm such that the flies are sterile. After being irradiated, the male flies are released to mate with females, but because the chromosomes of the sperm are damaged, the eggs never develop into viable larvae. Irradiating male flies was first successfully employed in the 1950s to control a blowfly, *Cochliomyia hominivorax*, that attacks grain crops.

Robber fly with green bottle fly prey

ROBBER FLIES, ASSASSIN FLIES

Latin name: *Asilidae* sp.

Family: Asilidae

Identification: 0.6 to 1 inch long, short spiny legs and spiny face, depression between large eyes

Distribution: Throughout North America, including tundra

Comments: Robber flies are ambush hunters. They are strong fliers with spines on their legs and faces that they use to hold their prey. They typically perch on plants, waiting for an unsuspecting insect to fly by. They then swoop down and capture the insect, hold on to it with their spiny faces and legs, and inject toxins that paralyze their prey and digestive enzymes that digest the organs of the insect. They then suck up the partially digested organs. Robber flies prey on almost any insect that comes by, including flies, bees, wasps, grasshoppers, beetles, dragonflies, and other robber flies. The larvae, which live in the earth, are also predators. Usually, they feed on the larvae of other insects.

Hairy face and forelegs of a devon red-legged robber fly

Robber fly on flower

ROBBER FLIES, BEE-LIKE ROBBER FLIES

Latin name: *Laphria* sp.

Family: Asilidae

Identification: 0.6 to 1 inch long, hairy, black and yellow, bee mimic

Distribution: Throughout North America

Comments: Robber flies of the genus *Laphria* are convincing bee and wasp mimics. The larger species mimic bumblebees, while the smaller species mimic honey bees or yellow jackets. There are two reasons for these fast-flying predatory flies to mimic bees. First, birds may think they are bees or wasps and leave them alone, as birds do not want to tussle with bees or wasps. Second, a bee may think that the robber fly is another bee and by the time it realizes that it is not, it is too late. When robber flies catch prey, they immediately bite it and deliver paralyzing toxins, which almost instantly leave the prey unable to move as it is gradually consumed by the robber fly.

ROBBER FLIES, GNAT-OGRES

Latin name: *Holcocephala* sp.

Family: Asilidae

Identification: 0.2 to 0.3 inch long, prominent divot between large eyes, large back feet

Distribution: Throughout North America

Comments: These small robber flies attack small flying insects, especially flies, although their diet depends on the insects that are available. They often perch on the tips of plant leaves on sunny days, waiting and watching for an unsuspecting insect to fly by. They attack their prey, injecting paralyzing toxins and digestive enzymes. They then usually return to their perch to suck up the liquid contents of their victim. Female flies lay their eggs in the soil in a mass that they cover with a chalky coating for protection. Like adult robber flies, the larvae are predacious, feeding on eggs, larvae, and other soft-bodied insects that live in the ground. The larvae or pupae overwinter and emerge as adults.

THREAD-WAISTED ROBBER FLIES

Latin name: *Leptogastrinae* sp.

Family: Asilidae

Identification: 0.3 to 0.6 inch long, long thin abdomen, large eyes, long back feet

Distribution: Throughout the United States and southern Ontario, absent from Pacific Northwest

Comments: *Leptogastrinae* are primitive robber flies. Unlike more-modern robber flies, which have a relatively short abdomen, these flies are characterized by a long abdomen. You might mistake one for a damselfly or a tiphiid wasp. They fly slowly with their long back legs dangling down, giving them a wasplike appearance. However, *Leptogastrinae* can put on a burst of speed when they are chasing prey insects, although they can't fly as fast as the more-modern robber flies.

Thread-waisted robber fly on a twig

STILETTO FLIES

Latin name: *Thereva* sp.

Family: Therevidae

Identification: 0.3 to 0.5 inch long, black with white vertical stripes on long tapering abdomen

Distribution: Throughout most of North America

Comments: Stiletto flies get their name because they have relatively long tapering abdomens, which must have reminded someone of the long tapering blades of stiletto knives. They are usually uncommon, but sometimes can be found on beaches and in meadows. These little flies can easily be mistaken for small robber flies but do not have the long spikes on their faces and forelegs that are characteristic of robber flies. Most species have one generation per year and overwinter as larvae. Adults feed primarily on nectar, honeydew, and pollen. Larvae live in soil or decaying wood, where they feed on fly, beetle, and moth larvae and other small soft-bodied invertebrates.

Female goldenrod gall fly

GOLDENROD GALL FLY, GOLDENROD BALL GALLMAKER

Latin name: *Eurosta solidaginis*

Family: Tephritidae

Identification: 0.2 inch long, brownish orange with black markings on wings

Distribution: Throughout eastern North America

Comments: Adult goldenrod gall flies emerge from their galls in the spring. Males perch on the goldenrod plant and wait for females to fly by. When males see females, they attract them by flicking their wings. After mating on the plant, females insert their ovipositor into the goldenrod flower buds and lay their eggs. When the larvae hatch, they crawl down the stem and secrete chemicals that induce the plant to form galls. The galls provide food for the larvae, protect them from predators, and shelter them from the rain, wind, and cold. Larvae overwinter in the gall. They synthesize sorbitol and glycerol, molecules which act as antifreeze to prevent their tissues from forming ice crystals. In the spring, the larvae pupate in the galls and emerge as adults through holes that they make in the galls.

Gall of a goldenrod gall fly

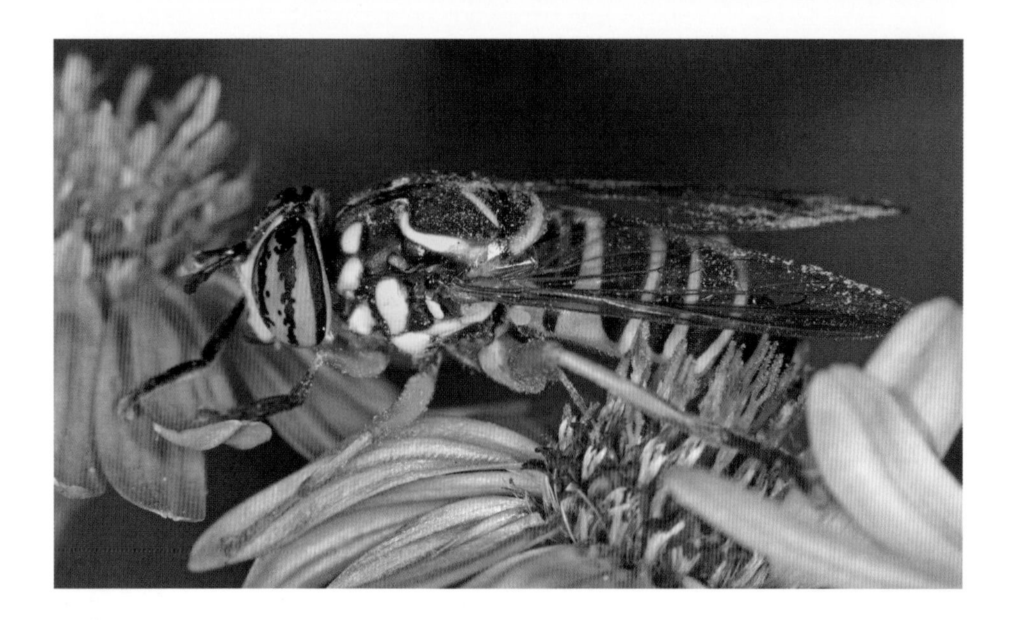

SYRPHID FLY

Latin name: *Spilomyia longicornis*

Family: Syrphidae

Identification: 0.7 to 0.9 inch long, long and slender abdomen with bright yellow and black stripes, yellow jacket wasp mimics

Distribution: Northeastern United States, west to Texas; southern Ontario

Comments: With their yellow and black stripes, these large hover flies are convincing yellow jacket mimics. They even extend their legs to mimic the antennae of yellow jackets. This kind of mimicry, where a harmless insect mimics a harmful insect, is called Batesian mimicry. It is directed at predators so they will think that the harmless insect is really the insect that it is mimicking. These large flies are important pollinators because they visit flowers. Flies are often the principal pollinators at higher altitudes because they are so robust that they can thrive at high altitudes where it is too cold for many other flying insects.

THICK-HEADED FLIES

Latin name: *Physocephala* sp.

Family: Conopidae

Identification: 0.3 to 0.4 inch long, black with white markings, wasplike narrow waist, head wider than thorax, long antennae that are enlarged at the tip, dark wings

Distribution: Throughout North America

Comments: Thick-headed flies of the genus *Physocephala* mimic potter wasps. Female flies are usually found on flowers, feeding on nectar and patiently waiting for a bee to arrive at the flower, or a nearby flower, or to fly by. When they see a bee, they take off, follow the bee, catch up with it, grab hold, insert their ovipositor into the soft exoskeleton between the bee's abdominal segments, and lay their egg. Sometimes they force the bee to the ground before they lay an egg in it. When the egg hatches, the larva feeds on the hemolymph of the bee before it consumes the bee's organs. After the bee dies from its wounds, the parasitoid fly larva pupates inside its dead host. The larvae of some species somehow induce their host to dig its grave into the ground before they pupate.

Thick-headed fly feeding on honeydew that was left on a leaf

THICK-HEADED FLIES

Latin name: *Myopa* sp.

Family: Conopidae

Identification: 0.15 to 0.25 inch long, brown with white markings on abdomen, white face on wide head

Distribution: Throughout North America

Comments: Like other flies in the family Conopidae, *Myopa* species are parasitoids. Usually, the hosts of their larvae are bees or wasps. Female flies chase their prey and lay an egg inside the bee or wasp. You might think a fly would be the loser if it tries to tangle with a bee or wasp, but these kinds of flies are faster than bees and more agile in the air, so they are able to catch bees, hold on to their abdomen, and lay an egg in them before the bee can fight back or escape. The abdomen of these flies is shaped somewhat like a can opener, so it can pry open the abdominal segments of a bee in order to insert an egg. The ovipositor also has sense organs that enable the fly to determine if is in the proper position to lay its egg.

FLESH FLIES

Latin name: *Sarcophaga* sp.

Family: Sarcophagidae

Identification: 0.2 to 0.9 inch long, black and gray longitudinal stripes on thorax, checkered pattern on abdomen, red eyes and setae on rear of abdomen

Distribution: Throughout North America

Comments: Flesh flies are different from most other flies in that the eggs hatch inside their mother and she produces live larvae instead of eggs. Animals that have their eggs develop inside their mother who then produces live young are referred to as ovoviviparous. Some kinds of fish, amphibians, reptiles, and invertebrates are also ovoviviparous. Adults and larvae of most species of *Sarcophaga* feed on carrion, dung, decaying fruit or vegetable matter, and in some species, open wounds of mammals. Flesh flies are often found in compost piles and pit latrines. Because larvae are born live and complete their development rapidly, flesh flies usually have several generations a year.

26 WASPS

The most important evolutionary advance for wasp ancestors was the development of social communities with three different castes: workers, males, and queens. Queens produce all the offspring; workers do the work, including expanding the nest, foraging for food, and feeding the larvae; and males mate with new queens. Since males do not build the nest, forage for food, or care for larvae, not many males are needed, and those that are needed are only required when new queens develop, usually in the fall.

It would be a waste of energy if half of the queen's eggs were to develop into males, as is the case with most other insects. Thus, queen wasps have a way to control the gender of their offspring. After mating, wasps and other insects store sperm. Sperm are released as eggs pass down the oviduct. If a queen wasp releases a spermatozoan when an egg passes down her oviduct, the fertilized egg will always develop into a female. However, if she doesn't release a spermatozoan, the unfertilized egg will develop into a male. In this way, the queen only produces males when they are needed to mate with new queens. Solitary wasps, bees, and ants also use the same method to control the gender of their offspring.

Many species of wasps are parasitoid. Parasitoids lay their eggs on or in another insect species, usually a particular kind of insect or group of closely related "host" insects. The eggs hatch on or in the insect host, and the larvae gradually devour it, being careful to avoid vital organs until they are ready to pupate. And what controls parasitoid insects? Well, parasitoid insects are often hosts of other parasitoids. It has been estimated that roughly 10 percent of insect species are parasitoids. When insect pests are accidentally imported, their numbers often increase rapidly because the parasitoid insect that controls the pest was not imported. When this happens, entomologist travel to the place where the pest originated and capture the pests that may be infected with a parasitoid. When the parasitoid leaves the pest, the scientists breed the parasitoid on the hosts to increase their numbers and bring the parasitoid back to North America to control the pests. In some cases this practice has been effective, though in other instances the imported parasitoid fails to become established or uses another insect species as a host.

Dusty birch sawfly caterpillar, *Craesus latitarsus*

COMMON SAWFLY LARVAE

Family: Tenthredinidae

Identification: Caterpillars with more than 5 pairs of prolegs

Distribution: Throughout North America

Comments: Not all caterpillars become butterflies or moths—some become sawflies or scorpionflies. Sawflies are primitive wasps, so named because they lay their eggs in slits in plant stems that they make with their sawlike ovipositor. It is believed that wasps, as well as butterflies and moths, may have evolved from insects that looked like sawflies or scorpionflies. Although they don't look like sawflies or scorpionflies today, butterflies and moths have kept the caterpillar-like larvae of their sawfly ancestors. At first glance, the larvae of most sawflies look like those of butterflies or moths. However, the caterpillars of butterflies and moths have a maximum of five pairs of prolegs, whereas the caterpillars of sawflies have more than five pairs of prolegs. Also, the prolegs of sawflies lack the minuscule spikes called crochets that are present on the prolegs of butterfly and moth caterpillars.

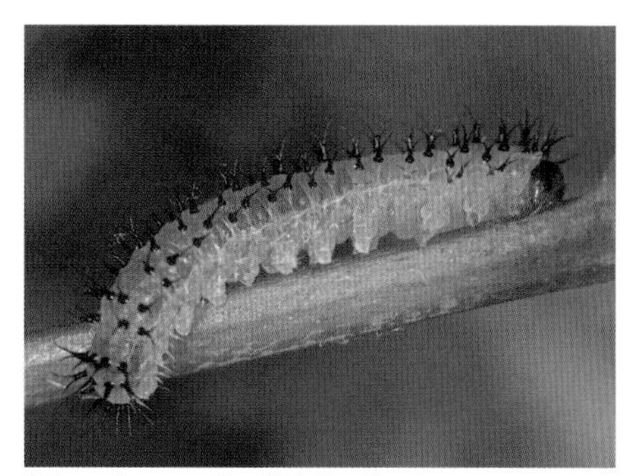

Oak sawfly caterpillar, *Periclista lineolate*

SAWFLIES

Latin name: *Nematus* sp.

Family: Tenthredinidae

Identification: 0.3 to 0.4 inch long, most species light tan, 6-segment antennae

Distribution: Throughout the United States and southern Canada

Comments: Adult sawflies usually only live for a week to 10 days. They feed on nectar and pollen and are minor pollinators. On the other hand, sawfly larvae live longer, and they feed on leaves, potentially causing significant damage to the plants and trees on which they feed. *Nematus* is a large genus of sawflies with sixty species in North American. The larvae of *Nematus ribesii* feed on the leaves of fruit bushes and trees and can cause major damage. The larvae of another species, the willow sawfly (*Nematus ventralis*), are common pests of willow and poplar trees. Although they usually do not cause serious damage to older trees, they can retard the growth of young trees. *Nematus ribesii,* the gooseberry sawfly or imported currantworm, is a serious pest of gooseberries and currants. Their caterpillars can completely defoliate those plants.

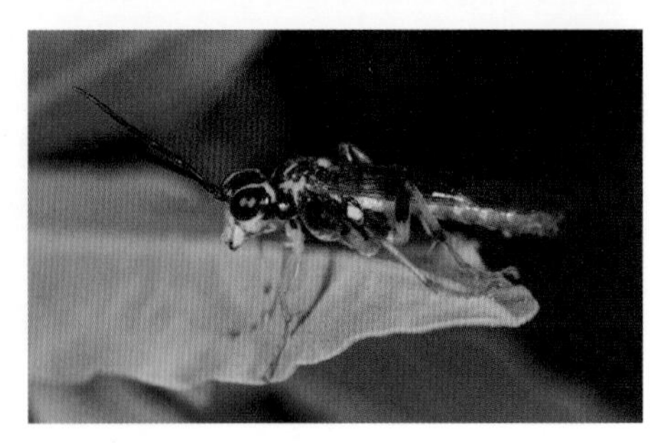

NORTHERN SAWFLY

Latin name: *Tenthredo originalis*

Family: Tenthredinidae

Identification: 0.3 to 0.4 inch long, wasplike, yellow legs with black upper sections, black antennae, smoky-colored wings

Distribution: New England and adjacent northeastern Canada, north to Labrador

Comments: Northern sawflies live in meadows and along streams, where they are often seen on flowers feeding on pollen and nectar. Although they are primitive wasps, sawflies do not sting, and predators know it. Therefore, some species like those of the genus *Tenthredo* are wasp mimics. Their black and yellow coloration and ovipositor that looks like a stinger make them convincing mimics of wasps that sting. However, unlike most wasps, northern sawflies are weak fliers. Like many other insects, the forewings and hindwings are attached by hooks so that the two wings on each side beat as one.

HORNTAIL, PIGEON HORNTAIL, PIGEON TREMEX

Latin name: *Tremex columba*

Family: Siricidae

Identification: 1 to 1.5 inches long, black with yellow stripes, dark wings, orange and black thorax, horn on abdomen

Distribution: Quebec, New Brunswick, Nova Scotia, and eastern United States, west to Colorado, Utah, and Arizona

Comments: Horntails are primitive wasps that are more closely related to sawflies than modern wasps. They do not care for their young or have a stinger. Although they look menacing, horntails are harmless. Females bore holes in trees, where they lay their eggs. They attack both deciduous and conifer trees, although horntails prefer maple, beech, hickory, and elm trees. Females insert their ovipositor about ¾ inch into a tree and lay an egg, and they keep boring holes until they have deposited 200 to 400 eggs. When the larvae hatch, they chew tunnels a foot or two long during the two or three years before they pupate and emerge as adults. Furniture that is constructed of wood that has not been kiln-dried may contain live horntail larvae that can hatch and frighten the homeowner who has not read this book and does not know that horntails are harmless.

BALD-FACED HORNET, WHITE-TAILED HORNET, BLACK JACKET

Latin name: *Dolichovespula maculata*

Family: Vespidae

Identification: 0.5 to 0.75 inch long, black with white markings on face, thorax, and abdomen

Distribution: Across southern Canada to Alaska and throughout United States

Comments: Although they are called hornets, bald-faced hornets are actually yellow jackets. The only true hornet in North America is the European hornet (*Vespa crabro*), which was introduced by settlers in the 1800s. Bald-faced hornets build large impressive hives, usually on branches of trees or bushes. In the fall, the nest is abandoned and all the wasps die except for the new queens, who overwinter in protected places like under bark or in hollow trees. New queens are fertile, having mated the previous summer. In the spring they chew wood that they mix with their saliva to make paper for the construction of the first few cells of a new nest. The queens lay eggs, and when the larvae hatch, they feed them chewed insects. When the adults emerge, they enlarge the nest, defend it, and feed the larvae. By midsummer, there may be 100 to 500 wasps in the colony. Like other yellow jackets, bald-faced hornets defend their nest. In addition to stinging, bald-faced hornets can squirt venom into the eyes of vertebrate intruders.

Bald-faced hornet nest

EASTERN YELLOW JACKET, EASTERN YELLOWJACKET

Latin name: *Vespula maculifrons*

Family: Vespidae

Identification: 0.4 to 0.65 inch long (smaller than similar species), black and yellow markings, black antennae

Distribution: Atlantic coast, west to North Dakota and Texas

Comments: Eastern yellow jackets are colonial wasps that build underground paper nests. Colonies consist of queens, workers, and males. New queens that have mated in the fall overwinter and emerge in the spring. Males and workers do not survive the winter. The new queens build small new paper nests, usually in an abandoned rodent burrow. They lay eggs in the new nest and forage for insects that they sting and bring back to feed their larvae. When the larvae develop into sterile female workers, they take over the jobs of enlarging the nest, caring for the larvae, and defending the nest. By the end of the summer, the nest may house hundreds or even thousands of individuals.

One of the important advantages that modern wasps like yellow jackets have over primitive wasps like sawflies and horntails is the thin "waist" on the front of their abdomen that acts as a hinge, allowing the wasps to turn around in the narrow passageways of their nests. This has made it possible for waisted wasps like yellow jackets to build narrow brooding chambers where they could not turn around if their waist could not bend.

Yellow jackets feed on nectar and tree sap and feed their larvae insects that they have captured, stung, and chewed. However, adults are also fond of fruit and meat, which is why they like to visit picnics and barbecues. If they come to your picnic, it is best not to try to harm them because they give a painful sting and can sting repeatedly. Yellow jackets also secrete alarm pheromones. Although nearby yellow jackets usually do not respond to the pheromones, workers in the nest quickly fly to the aid of their sister. If a yellow jacket visits your picnic, my suggestion is to enjoy watching it walking around the picnic table as it moves its antennae to explore its surroundings. Yellow jackets are usually so involved in searching for or eating their lunch that they will not notice you even if you get very close to them.

EUROPEAN PAPER WASP

Latin name: *Polistes dominula*

Family: Vespidae

Identification: 0.5 to 0.6 inch long; black and yellow lines on head, thorax, and abdomen; yellow antennae

Distribution: Throughout North America

Comments: Although paper wasps look like yellow jackets, they can be distinguished by their mostly tan antennae, whereas yellow jackets have black antennae. *Polistes dominula* nests are constructed of paper composed of a mixture of chewed wood and saliva. The European paper wasp was first discovered in Cambridge, Massachusetts, in the late 1970s. Since then the species has been found in almost every state and Canadian province. In many areas it has displaced the native North American species of paper wasps. A number of reasons account for the success of the European paper wasp in North America. First, they have a shorter generation time than other paper wasps, so they can expand the number of wasps in their nest quicker. Also, most paper wasps feed their larvae caterpillars; however, *Polistes dominula* feed their larvae other insects in addition to caterpillars. This species also builds their colonial nests in protected areas. In addition, European paper wasps are particularly formidable wasps.

Nest of the European paper wasp

Female mason wasp on her nest

MASON WASP, POTTER WASP

Latin name: *Ancistrocerus catskill*

Family: Vespidae

Identification: 0.4 to 0.6 inch long; black; 4 yellow stripes on back of abdomen, 2 on front of abdomen

Distribution: Eastern United States, southeastern Canadian provinces

Comments: Mason wasps get their name because some species build their nests out of mud, although other species nest in crevices in bark or rocks. Potter wasps get their name because the nests of some species resemble pots, whereas the nests of other species do not. The names, potter wasp and mason wasp, are often used synonymously. Female *Ancistrocerus catskill* construct mud nests, where they raise their larvae. The nest is divided into individual chambers, and the wasp lays a single egg in each chamber. She then captures and stings caterpillars, beetle larvae, and spiders, which renders them paralyzed for the rest of their short lives, and brings them home to feed her larvae.

POTTER WASP

Latin name: *Monobia quadridens*

Family: Vespidae

Identification: 0.7 inch long; black; large white stripe on front of abdomen, small white stripe on back of thorax

Distribution: Southern Ontario, throughout eastern United States, west to Wisconsin and New Mexico

Comments: Many potter wasp species get their name because they make earthen nests that look like pots. However, some species of potter wasps, like *Monobia quadridens*, nest in preexisting cavities such as beetle tunnels in dead wood or abandoned wasp nests. Females partition their nest into separate chambers, one for each of their eggs. Before laying their eggs, the female potter wasps capture and sting caterpillars that will be used for food for their larvae. Larvae feed on the caterpillars for about a week before they pupate. *Monobia quadridens* has two generations a year, with the last generation of the summer overwintering as pupae. The females are reluctant to sting, but they are capable of delivering a painful sting. Unlike most other wasps, males can also sting. They have no stinger or venom, but they have a sharp tip on their abdomen. Since they have no venom, the sting is transient, but it might frighten off a predator.

POTTER WASP

Latin name: *Eumenes fraternus*

Family: Vespidae

Identification: 0.5 to 0.6 inch long, black with ivory-colored markings, first abdominal segment thin at front and widening at back

Distribution: Eastern United States and Ontario

Comments: *Eumenes fraternus* feed their larvae caterpillars. Females build pitcher-shaped nests for each of their young. To construct the nests, the wasps collect a drop of water. They then collect some dry soil, dampen the soil, and place it in a precise locale to begin constructing a tiny pot. They usually need to repeat this several hundred times before the pot is completed. When the pot is finished, the females lay an egg in the pot. They then search for an insect, usually a caterpillar, sting it, return to the pot, and place the paralyzed caterpillar in the pot for their larva to feed on when it hatches. If the caterpillar isn't big enough, they find and add another one. When a female is satisfied that her larva will have enough to eat, she seals the pot with mud and starts work on the next pot. Larvae mature, pupate, chew a hole in the pot, and crawl out into the world as adults.

SAND WASP

Latin name: *Bicyrtes quadrifasciatus*

Family: Crabronidae

Identification: 0.6 inch long, black with yellow markings on head and thorax, yellow stripes on abdomen, black antennae and eyes, yellow legs

Distribution: Eastern United States and southeastern Canada

Comments: Sand wasps dig their nests in sand or loose soil. Although they are solitary wasps, several females often dig their nests close together. They also often gather in groups to fight off intruders. Sand wasps feed on nectar, but like other large wasps, they feed their young insects that the mothers capture, paralyze by stinging, and carry back to stock the nest for their larvae. The nests are simple narrow tunnels with a chamber at the bottom. The good news is that wasps of the genus *Bicyrtes* feed their young stink bugs and leaf-footed bugs that are pests in gardens and on farms. The bad news is that sand wasps often dig their nests in sandboxes. Although, like most solitary wasps, they do not sting unless they are harmed, they may sting a child who inadvertently injures them while digging in the sandbox.

Beware of sand wasps in the sandbox.

WEEVIL-HUNTING WASPS

Latin name: *Cerceris* sp.

Family: Crabronidae

Identification: 0.4 to 0.6 inch long, large head with curved projections on their face, some species black with yellow markings

Distribution: Throughout North America

Comments: Members of the genus *Cerceris* feed their larvae weevils or bees. Females dig nests in the ground. They prefer the sandy soil along roadsides, baseball fields, parks, and beaches. The wasps begin their nests by digging a hole, then compact the soil to create a chamber where they lay an egg. Although they are solitary wasps, they usually build their nests adjacent to the nests of other weevil-hunting wasps of their species. After laying an egg, the wasps fly off and search for a weevil. Once they find a suitable prey, they alight on it, grab it by the thorax with their mandibles, and insert their stinger at the base of the weevil's leg, where there is soft chitin that is easy to penetrate. They then hold the paralyzed beetle with their long, sharp mandibles, carry it back to their nest, and place it on the egg for their larva to eat when it hatches, before beginning work on another chamber.

BEE WOLVES, BEE HUNTERS, BEE KILLERS

Latin name: *Philanthus* sp.

Family: Crabronidae

Identification: 0.4 to 0.6 inch long, black with yellow stripes on abdomen and yellow markings on head, large eyes

Distribution: Throughout North America

Comments: Members of the genus *Philanthus* are solitary wasps that feed on nectar and pollen. They feed their larvae bees that the females intercept in midair and paralyze by stinging them on their underside, before bringing them back to their nests. A European species, *Philanthus triangulum*, feed their larvae honey bees, which can be a problem for beekeepers, but North American species of bee wolves feed their larvae a variety of bee species. Bee wolves nest in tunnels in the ground that are dug by females. The tunnels can be 2 or 3 feet long, with a number of individual brood chambers off of the main tunnel. Male bee wolves mark the site with twigs or leaves and release pheromones to deter other males from the nest.

NO COMMON NAME

Latin name: *Eupelmus vesicularis*

Family: Eupelmidae

Identification: 0.08 inch long, brown and black, wingless

Distribution: Throughout North America

Comments: This minute wasp lives in leaf litter, where it is a parasitoid to at least 200 insect species. Eupelmidae is one of several families that belong to a larger "superfamily," the Chalcidoidea. Wasps in the superfamily are referred to as chalcids. They are mostly tiny wasps and often overlooked by insect enthusiasts because, like *Eupelmus vesicularis*, they are so small. However, chalcids are extremely numerous and have a tremendous impact on our lives because they are parasitoids of most, if not all, insect pests. It is estimated that worldwide there are more than 500,000 species of chalcid wasps, the large majority of which have not been described. I don't know how the estimate was made, especially since the great majority of these mostly minuscule insects have not been described. Sometimes statements in the literature are quoted many times. They may or may not be accurate.

CHALCID WASPS

Latin name: *Perilampus* sp.

Family: Chalcidoidea

Identification: 0.1 to 0.3 inch long, iridescent black, large modeled (bumpy) thorax

Distribution: Throughout North America

Comments: Most chalcids are tiny parasitoid wasps. The chalcid wasp *Perilampus* is a parasitoid of the larvae of other parasitoids. Female chalcid wasps usually lay their eggs on flowers. When the eggs hatch, the larvae wait for a caterpillar. These larvae can survive for long periods of time without eating. When a caterpillar eventually comes along, they attach themselves and penetrate the caterpillar. If there are no larvae from another species of parasitoids, the *Perilampus* larvae do not develop. However, if there are parasitoid larvae inside the caterpillar, *Perilampus* larvae wait until the caterpillar pupates. When it does, the *Perilampus* larvae attack and eat the parasitoid larvae.

TIPHIID WASPS

Latin name: *Myzinum* sp.

Family: Tiphiidae

Identification: 0.3 to 0.7 inch long, abdomen alternating yellow and black bands

Distribution: Throughout the United States and southern Canada

Comments: Adult tiphiid wasps feed on the nectar of a number of plants, especially plants of the aster and allspice families. Their larvae are parasitoids of scarab and tiger beetles. Female wasps search out the burrows that are made by female beetles and follow the tunnels to find a larva. When the female wasp finds a larva, she lays her eggs on its abdomen and stings the larva, rendering it partially paralyzed. The wasp larvae feed on the beetle larva until it is consumed and then pupate next to the dead beetle larva, overwinter as pupae, and emerge as adult tiphiid wasps in the spring.

THREAD-WAISTED WASP, HUNTING WASP

Latin name: *Ammophila procera*

Family: Sphecidae

Identification: 0.7 to 2 inches long, long thin waist, black with orange segment in abdomen

Distribution: Throughout North America

Comments: Thread-waisted wasps feed on nectar, but they feed their larvae insects. Female *Ammophila procera,* and some other species of thread-waisted wasps, dig a nest as a hole in the ground. They then capture, sting, and permanently paralyze a caterpillar, place it in their nest, lay an egg on the caterpillar, and cover the hole so that it will not be seen by predators. Thread-waisted wasps repeat this process many times until all of their eggs are laid, each in a different nest. These wasps remember where their nests are and regularly visit each one, uncovering it to monitor if their larva has consumed the paralyzed caterpillar. If so, they re-cover the hole, find and paralyze another caterpillar, return to the nest, uncover the hole, and feed the caterpillar to the larva, before re-covering the hole again. Who says insects can't learn and remember? I sometimes can't remember where I parked my car in the parking lot.

GRASS-CARRYING WASPS

Latin name: *Isodontia* sp.

Family: Sphecidae

Identification: 0.7 to 0.8 inch long, black, hairy thorax, brown wings

Distribution: Eastern North America, west to the Rocky Mountains

Comments: The common species of grass-carrying wasp, *Isodontia mexicana,* was accidentally introduced to Europe from North America. These wasps are thriving in Europe, probably due to the lack of natural predators and parasitoids. Grass-carrying wasps build their nests in natural cavities such as hollow branches or abandoned beetle tunnels and line their nests with grass. When they are flying, they are often seen carrying a long blade of grass. At first glance, they appear to have a long, thin tail. Before laying their eggs, the wasps catch and sting grasshoppers, tree crickets, or katydids, which they bring home as provision for their nests. They then lay their eggs and the helpless prey are gradually eaten by the grass-carrying wasp larvae.

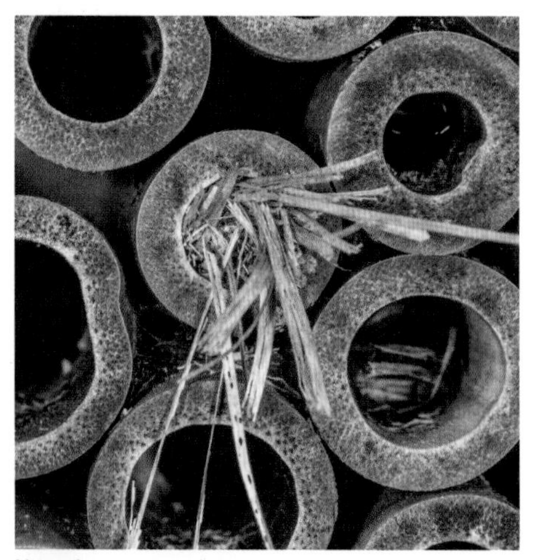

Nest of a grass-carrying wasp

Being eaten alive seems like a terrible fate to us. However, it is the fate of most insects, and it is assumed that they do not feel pain as we do. In fact, no pain receptors have been found in insects.

ECTEMNIUS

Latin name: *Ectemnius* sp.

Family: Sphecidae

Identification: 0.2 to 0.3 inch long, large head and eyes, black with white markings on head, thorax, and legs

Distribution: Eastern North America

Comments: Some species of *Ectemnius* nest in the pith of twigs that they partition into several cells. Other species excavate nest tunnels in soft wood like rotten logs, tree stumps, or fallen trees. Females feed their larvae flies that they catch, sting, and bring back to the nest. *Ectemnius* is a solitary wasp. Unlike colonial wasps that build nests of numerous individuals and have queens, males, and workers that care for all the young, female solitary wasps make their own nest and take care of their own young. However, some species of *Ectemnius* and other solitary wasps build nests that house several females. Unlike colonial wasps, which vigorously defend their nests, solitary wasps do not defend their nests. Also, they are not aggressive and are less willing to sting than colonial wasps.

ENSIGN WASPS, NIGHTSHADE WASPS, HATCHET WASPS

Latin name: *Evania* sp.

Family: Evaniidea

Identification: 0.4 to 0.6 inch long, black, slender, abdomen resembles a flag

Distribution: Throughout North America, wherever cockroaches are found

Comments: You have to love ensign wasps because they lay their eggs in the eggs of cockroaches. Adults drink nectar from flowers, but their larvae feed on and eventually kill cockroach eggs and embryos. They are called ensign wasps because their abdomen resembles a flag (an ensign is the country flag on a ship). After mating, female ensign wasps search out a female cockroach that is carrying an egg case and lays her eggs inside the cockroach eggs. The ensign wasp eggs quickly hatch and consume the cockroach embryo. Some species have a worldwide distribution, having traveled with cockroaches. The best place to find these little wasps is in buildings that are infested with cockroaches. However, they are never in large enough numbers to control cockroach populations.

CUCKOO WASPS, GOLD WASPS

Latin name: *Chrysis* sp.

Family: Chrysididae

Identification: 0.5 inch long, brilliant green, mottled body

Distribution: Throughout the United States and southern Canada

Comments: Some species of cuckoo birds lay their eggs in the nests of others when it is unattended. When the female bird returns, she assumes the eggs are hers and raises the offspring. This kind of behavior is called kleptoparasitism. Cowbirds exhibit the same behavior, which is why many birders are not very fond of cowbirds. Like their bird namesake, cuckoo wasps lay their eggs in the nests of bees and other wasp species. Some species remove the eggs of the host wasp and lay their own eggs in the nest. Similar to the cuckoo birds, when the female returns, she raises the cuckoo wasp larvae as if they were her own. Other species of cuckoo wasps kill the female or enslave her to raise their larvae. There are cuckoo bees that exhibit the same behaviors.

Female velvet ant

VELVET ANT

Latin name: *Pseudomethocha* sp.

Family: Mutillidae

Identification: 0.6 to 1 inch long; wingless females bright orange and hairy, with black antennae and legs; winged males black with orange area on abdomen

Distribution: Eastern North America

Comments: Although they are called ants, velvet ants are wasps. Males are usually dark-colored and have wings. Females are wingless, hairy, and bright red or orange. On sunny days, female velvet ants walk around looking for bee nests. Upon finding one, they enter the nest and lay their eggs on the bee larvae. When the eggs hatch, the velvet ant larvae consume the bee larvae as they grow. They pupate and emerge as adults in the bee's nest. Although they are often out in the open, the flightless female velvet ants don't have that much to fear. In addition to their bright red or orange warning colors, they have an extremely hard exoskeleton and an exceptionally painful sting. Predators avoid them, and you would be wise to do the same.

Male velvet ant

OAK ROUGH BULLET GALL WASP

Latin name: *Disholcaspis quercusmamma*

Family: Cynipidae

Identification: 0.05 to 0.08 inch long, reddish brown, chubby, round abdomen

Distribution: Throughout North America, wherever oak trees grow

Comments: The tiny oak rough bullet gall wasp is a common gall-producing species. Like many related gall wasps, they produce alternating sexual and parthenogenetic generations. Galls are growths that plants form in reaction to chemicals produced by certain insects, mites, fungi, viruses, or slime molds. They are rich in food and provide a protective home for insect larvae. Although insects produce galls in many plants, oak galls are the most common. Sometimes by late summer a number of other insects also decide that an oak gall would be a fine place to lay their eggs. Thus, the larvae of several kinds of insects, in addition to those of the gall wasp, can often be found inside the gall. Galls are somewhat similar to burls, the beautiful wood that is sometimes used in fine furniture and wooden objects. Like galls, burls are formed by the reaction of a tree to insect, bacterial, or fungal toxins.

ICHNEUMON WASPS

Latin name: *Rhyssa* sp.

Family: Ichneumonidae

Identification: 0.4 to 0.8 inch long, thin, black, females have long ovipositor

Distribution: Throughout North America

Comments: Ichneumon wasps have some of the longest ovipositors of any insect. Females of some species wander over logs, listening for the sound of larvae crawling in the log. Different species of ichneumon wasps seek the larvae of different insects, but those that lay their eggs in beetle grubs that live in logs have the uncanny ability to locate the exact position of a grub in the log. When they have located a larva, female wasps insert their long ovipositor through the wood, into the beetle grub, and lay their eggs. Sensory receptors on the ovipositor enable the wasp to determine when its ovipositor is inserted into the larvae. The eggs that they lay hatch into parasitoids that gradually consume the larvae as they grow, molt, and eventually crawl from the nearly dead larvae and pupate. Males also wander over the same logs, waiting for virgin females to crawl from the log.

ICHNEUMON WASPS

Latin name: *Amblyteles* sp.

Family: Ichneumonidae

Identification: 0.4 to 0.6 inch long, yellow abdomen with black bands on some species, yellow or yellow and black legs, females lack long ovipositor

Distribution: Throughout North America

Comments: Wasps of the genus *Amblyteles* belong to the Ichneumonidae family, which is the largest family of wasps with over 3,300 described species in North America. Many more species have not yet been described and may never be as they become extinct due to habitat loss and insecticides. Although many species of ichneumon wasps have long, menacing-looking ovipositors, the ovipositors of *Amblyteles* are used exclusively to lay eggs. None of the over 3,300 species of ichneumon wasps can sting. It is helpful to know which kinds of wasps sting so that they can be avoided. Also, you can impress people by fearlessly picking up an ichneumon wasp, horntail, or other kinds of wasp that cannot sting. *Amblyteles* species feed on nectar and pollen. Females lay their eggs on moth caterpillars, and when the eggs hatch, the larvae feed on the caterpillar.

SHORT-TAILED ICHNEUMON WASPS

Latin name: *Ophion* sp.

Family: Ichneumonidae

Identification: 0.3 to 0.7 inch long, light brown or reddish-brown body flattened side to side, long antennae and legs, 3 prominent ocelli

Distribution: Throughout the United States and southern Canada

Comments: Short-tailed ichneumon wasps are solitary wasps that feed on pollen and nectar. Like other ichneumon wasps, their larvae are parasitoids. The larvae of some species are parasitoids of various beetle larvae, while others are parasitoids of caterpillars, especially the caterpillars of giant silk moths, owlet moths, and tiger moths. Short-tailed ichneumon wasps breed during the spring and summer. Inseminated females lay one egg on their larval host. When the eggs hatch, the wasp larvae bore into the host and gradually consume it, being careful not to eat any vital organs until they are ready to pupate. Larvae pupate inside the host and emerge as adults. Hosts do not survive.

BRACONID WASPS

Latin name: *Atanycolus* sp.

Family: Braconidae

Identification: 0.2 to 0.3 inch long, black head and thorax, red abdomen, females have long ovipositor

Distribution: Throughout North America

Comments: Braconidae is a large family of wasps, with almost 2,000 described species in North America and probably many more that have not yet been described. Although they look like ichneumon wasps, they can be distinguished by their first and second abdominal segments, which are fused. All the wasps in this family are parasitoids of various insect species. Many species of the genus *Atanycolus* are parasitoids of caterpillars. If you see a caterpillar that has what appear to be eggs on its back, those are not eggs but cocoons of braconid wasp pupae. Because their larvae are parasitoids of insects, some species of braconids are raised for control of agricultural pests.

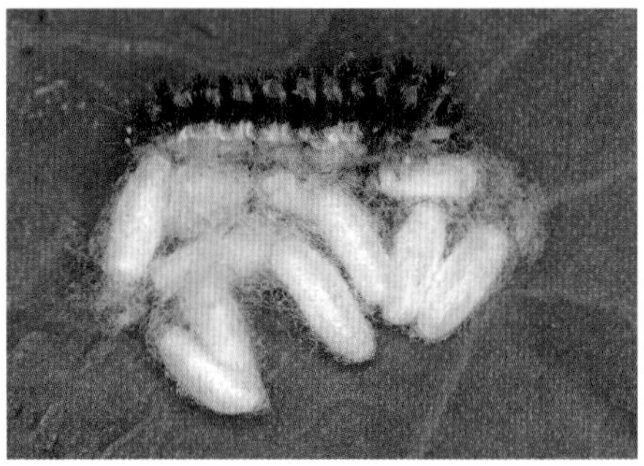

Parasitoid wasp pupae next to their dead caterpillar host

27
BEES

Bees evolved from wasps of the family Crabronidae about one hundred million years ago at the same time that flowering plants first appeared on Earth. The simultaneous evolution of flowering plants and bees is referred to as coevolution because flowering plants need insects to transport pollen from one flower to another and bees need pollen and nectar to feed their larvae. Unlike wasps, which raise their larvae on insects, bees feed their larvae pollen and nectar. Insect larvae need protein in order to thrive and grow. Wasp larvae get their protein from the insects that they are fed, and bee larvae get their protein from pollen.

In order to attract bees, flowers evolved bright visible colors, as well as ultraviolet colors, which bees see very well. Flowering plants also evolved the ability to make nectar containing sugar and scents to attract bees. At the same time, bees evolved long tongues to reach into flowers where nectar is stored and special hairs, called scopal hairs, to which pollen readily attaches. Today, bees remain the chief pollinator of flowering plants, although flies, butterflies, wasps, and other insects also pollinate plants. Most species of bees are solitary, but a few are colonial.

HONEY BEE

Latin name: *Apis mellifera*

Family: Apidae

Identification: Workers 0.4 to 0.8 inch long, golden brown and black with pale yellow bands on abdomen, long hairs and often pollen on hind legs

Distribution: Worldwide

Comments: *Apis mellifera* was introduced to North America by colonists in Jamestown, Virginia, in 1622. Honey bees were very important to the early settlers because they produced honey and beeswax and pollinated crops. Today, beekeeping, or apiculture, is a multimillion-dollar industry in North America. Beekeepers truck their beehives to farms, where they are placed so that the bees can pollinate almonds, apples, alfalfa, and other crops.

Honey bees are highly social insects with different castes: sterile female workers take care of and protect the hive and the young, as well as gather nectar and pollen that they make into honey; a queen produces all the offspring; and males (drones) fertilize new queens. Young workers take care of the eggs and larvae while older workers, which are more dispensable because they don't have as long to live, take on the more dangerous task of foraging for pollen and nectar. When a worker bee finds a good source of honey, she returns to the hive and does a little "dance" that communicates to her sisters the location of the flowers where she found the nectar and pollen.

Honey bees have a unique way of overwintering. They store up honey that they made from pollen and nectar in the spring, summer, and fall. When the temperature drops, some bees flap their wings, circulating

Honey bee gathering pollen and nectar. Note orange pollen on the bee's pollen basket (hind leg).

air in the hive and producing heat. When they tire, they take a break and feed on some honey, and other workers take their place.

Most honey bees live in man-made hives, but wild bees make their hives in hollow trees. When a colony of bees becomes too crowded, the queen lays eggs that become fertile males and females. When they pupate, they fly off, mate, and begin a new colony. When a honey bee colony is attacked, workers rush to defend their hive and their queen. Unlike wasps, honey bees can only sting once, because when they sting the base of the stinger detaches, which kills the bee.

Unfortunately, honey bees are on the decline because of mites that live in beehives, where they infect and kill bees. The increase in the use of insecticides is also taking a large toll on bees, as well as other beneficial insects.

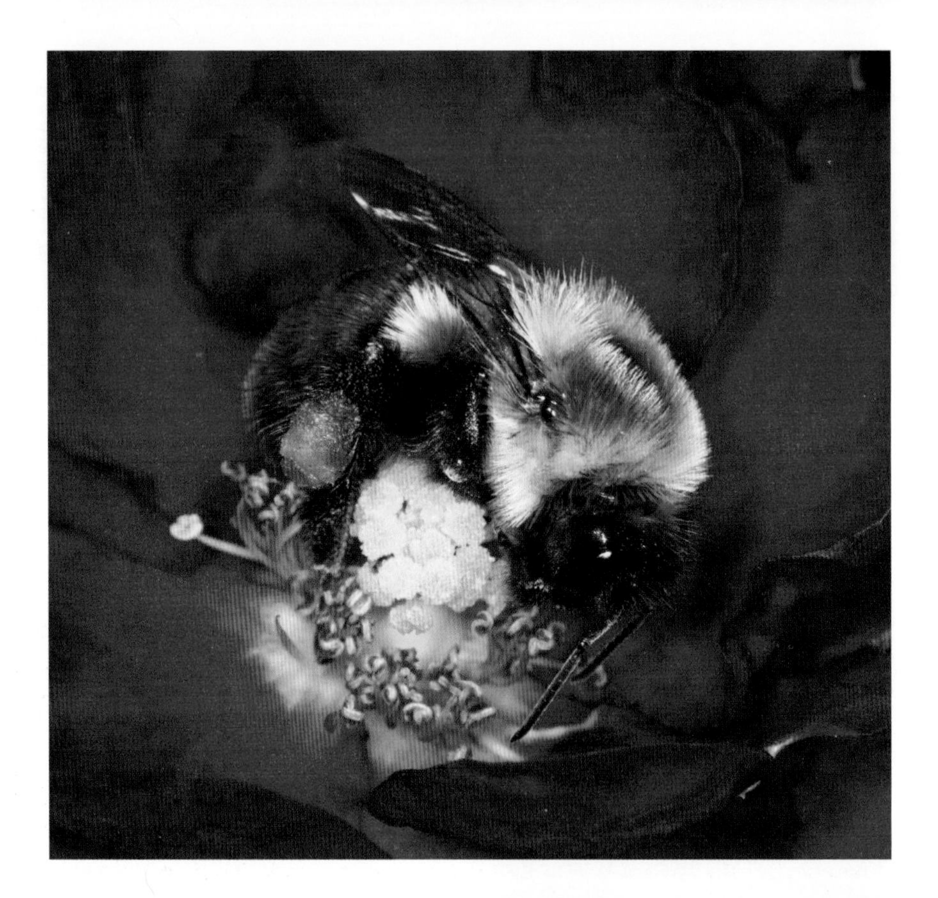

BUMBLEBEES, BUMBLE BEES

Latin name: *Bombus* sp.

Family: Apidae

Identification: 0.7 to 1.2 inches long, round bodies covered with branched setae, black and yellow bands

Distribution: Throughout North America

Comments: Bumblebees prefer to use abandoned underground tunnels that were dug by mice, rats, chipmunks, or rabbits rather than dig their own nests. Small colonies are established by a single queen. Bumblebees visit flowers to feed on nectar and carry it back to their nest as a food store. They also bring pollen back to their nest to feed their larvae. Pollen is attached to hairs on their rear legs, collectively termed a pollen basket.

Bumble bee using its long tongue to feed on nectar

Bumblebees are raised in large numbers to pollinate greenhouse plants. In Europe, greenhouses are an especially economical way to farm because land is costly. Bumblebees are the preferred pollinators in greenhouses because, unlike honey bees, they do not fly into glass. Also, bumblebees are more efficient pollinators than honey bees because they are active throughout the day and on cloudy days. Additionally, they do not require a hive because they live in the ground and they are less aggressive than honey bees.

EASTERN CARPENTER BEE

Latin name: *Xylocopa virginica*

Family: Xylocopinae

Identification: 0.5 to 1.2 inches long, similar to bumblebees but have shiny black abdomen

Distribution: Maine to Florida, west to Nebraska and eastern Texas

Comments: Eastern carpenter bees can be mistaken for bumblebees, as they are the same shape and size and have a similar color pattern, but they can usually be distinguished by their shiny black abdomen. Carpenter bees excavate nests in wood such as tree stumps, fence posts, and buildings, but they prefer milled pine or cedar, so they often build their nests in the wooden eaves of garages or barns, where they can cause some harm. However, the damage is usually superficial, especially because they do not eat wood. A nest may house a single female or several females. The tunnels that female carpenter bees construct in wood can be several inches long. The females partition the tunnels into several cells separated by cemented wood chips, then lay an egg in each cell and stock it with pollen and nectar on which the larva will feed. The larvae grow and spend the winter in their cells. In the spring, they pupate and emerge as adults.

LEAFCUTTER BEES

Latin name: *Megachile* sp.

Family: Megachilidae

Identification: 0.3 inch long, black with white setae between abdominal segments

Distribution: Throughout North America

Comments: Females leafcutter bees dig nests and line them with pieces of leaves that they cut with their sharp mandibles. Their nests are usually composed of a long columns of cells. The female bees lay an egg in each cell, stock the cell with pollen and nectar, and build caps to walls off each cell. When the larvae hatch, they consume the food as they grow, spin a cocoon, and pupate. They overwinter as pupae and emerge as adult bees in the summer. Leafcutter bees carry pollen on hairs beneath their abdomen and are excellent pollinators. One species, *Megachile rotundata,* is such a fine pollinator that it is raised on farms to pollinate crops, especially alfalfa.

MASON BEES

Latin name: *Osmia* sp.

Family: Megachilidae

Identification: 0.3 to 0.6 inch long, stout

Distribution: Throughout North America

Comments: There are more than forty-five species of the genus *Osmia* in North America. Some are native, while others have been imported. In the spring, male bees emerge from their nests and wait for females to appear. After mating, the males die, and the females search for a place to build nests for their young. They usually choose holes in wood that were made by woodpeckers or the abandoned nests of wood-boring beetles. Females create individual brood cells and supply them with pollen and nectar. They then lay eggs in the cells and seal them over with mud. When the eggs hatch, the larvae feed on the stores of pollen and nectar, mature, pupate, overwinter, and exit the chamber as adults in the spring.

Mason bees are often kept to pollinate gardens and farm crops. In addition to being excellent pollinators, female mason bees are reluctant to sting. They will also nest in man-made houses constructed of short lengths of bamboo or holes in wood. Both nests and bees are for sale on a number of websites.

RESIN BEE

Latin name: *Anthidiellum notatum*

Family: Megachilidae

Identification: 0.3 to 0.4 inch long, chunky, 4 rows of yellow spots on abdomen

Distribution: Throughout the United States and southern Ontario

Comments: Resin bees are stocky bees that belong to a large family of bees called Megachilidae, which includes mason bees, leafcutter bees, and others. They all carry pollen on scopal hairs that are located on the underside of their abdomen. Resin bees are solitary bees that build brooding chambers for their young from resin and pebbles. They use their mandibles to obtain resin and sap from pine trees, which they mix with pebbles to precisely construct brooding chambers. These brooding chambers are usually built under rocks, and each chamber accommodates a single egg.

MINING BEES, DIGGER BEES

Latin name: *Andrena* sp.

Family: Andrenidae

Identification: 0.3 to 0.6 inch long, pubescent, females larger and fatter than males

Distribution: Throughout southern Canada and the United States

Comments: As the name implies, mining bees excavate underground nests. They often hide the opening of their nest by placing a pebble, leaf, or stick over it. One individual usually stands guard while the others are off collecting nectar and pollen to feed the larvae. They are considered subsocial insects because they live in groups where each bee raises her own young. With social insects, only the queen is fertile, and workers take care of and feed all the larvae. Many kinds of bees and wasps dig underground nests where they raise their young. With all those underground nests of insects that can sting, it is a good idea to make sure that your kids have shoes on when they are playing in the yard.

SWEAT BEES

Latin name: *Augochlorella* sp.

Family: Halictidae

Identification: 0.3 to 0.5 inch long, bright green or copper

Distribution: Southern Quebec to Florida, west to Minnesota and Texas

Comments: Sweat bees get their name because on hot summer days they often alight on people in order to ingest sweat for the salt. Species in the genus *Augochlorella* are usually found on flowers in late winter and early spring. These solitary bees nest in the ground, where they line their brood cells with water-resistant molecules. Females fashion the pollen and nectar that they gather into a ball on which they lay their eggs. After they mate in the fall, the males die and females overwinter in their underground nests. *Augochlorella* species, as well as other sweat bees, are important pollinators of many garden and crop plants, especially apples, alfalfa, and sunflowers. If you would like these little bees to pollinate your garden, plant wildflowers and avoid insecticides.

CUCKOO BEES, SWEAT BEES

Latin name: *Sphecodes* sp.

Family: Halictidae

Identification: 0.3 to 0.5 inch long, wasplike, black head and thorax, red or red and black abdomen

Distribution: Throughout North America

Comments: Female sweat bees in the genus *Sphecodes* have found a clever, albeit diabolical, way to avoid the laborious task of collecting pollen and nectar and flying back and forth from flowers to their nest to feed their larvae. Collecting pollen and nectar is not only laborious but also dangerous, because there are many predators that capture and eat flying insects. So these bees walk around looking for a nest of other sweat bee species, then wait until the female flies off to gather pollen and nectar. When she is away, they enter her nest, destroy or remove her eggs, lay their own eggs in the nest, and seal it over. When their eggs hatch, the larvae feed on the food stores in the nest.

VIRESCENT GREEN METALLIC HALICTID BEE

Latin name: *Agapostemon virescens*

Family: Halictidae

Identification: 0.4 to 0.5 inch long, bright green head and thorax, brown abdomen with yellow stripes, females have pollen brush on legs

Distribution: Northeastern United States and Quebec, south to Florida, west to Texas and Oregon and British Columbia

Comments: These communal sweat bees nest in the ground, with several female bees sharing the same nest. Although the nest has only one entrance, each bee digs her own brood cell off a common passageway. She collects pollen and nectar that she fashions into balls, then places the balls in her brood cell and lays an egg on each one. Unlike social bees, there is no queen or workers, and the bees do not share duties. However, because there are several bees in the nest, it is easier for them to defend the nest from predators.

28
ANTS

Ants have been studied extensively, and there are some excellent books devoted to identifying ants or the biology of these interesting little colonial insects. It is estimated that if we were to weigh all the ants in the world and all the other land animals in the world, including people, the combined weight of the ants would probably be greater than the combined weight of the other land animals.

The most distinguishing characteristics of ants are their elbowed antennae and the distinctive node-like structure between the thorax and abdomen called the pedicel. Although it is usually easy to identify an insect as an ant, it is not very easy to determine the species of ant without a key, detailed drawings, and preferably a dissecting microscope, although a good 10x to 15x hand lens is usually sufficient. The ant has to be dead so that you can view it from different angles in order to see the details of its anatomy. If you are not used to using a key to identify insects, it can be frustrating, but it can also be satisfying when you are able to identify the species or genus of an ant.

DRACULA ANT

Latin name: *Amblyopone pallipes, Stigmatomma pallipes*

Family: Formicidae

Identification: 0.3 inch long, reddish brown, large mandibles, segmented abdomen

Distribution: Throughout North America

Comments: This ant has two Latin names. Experts frequently change the genus and species of an insect based on new information that suggests that another genus and species is more appropriate. They also frequently reassign insects to a different family. Thus the Latin name of an insect or what family it belongs to may be different in an older book than in newer references.

Dracula ants are primitive ants. They are rarely seen because they are small and do not leave the soil. However, they can be found by spreading soil on a white piece of paper or cardboard and moving the soil around with a small paintbrush. These little ants live in small colonies of only about twenty individuals. They feed almost exclusively on small centipedes that live in the ground and are unusual because instead of bringing their prey back to their nest like most ant species, they carry their larvae to the prey.

An ant takes a drop of honeydew from a treehopper. Ants protect treehoppers, aphids, and certain other hemipterans in exchange for sweet honeydew.

WOOD ANTS, MOUND ANTS, FIELD ANTS

Latin name: *Formica* sp.

Family: Formicidae

Identification: 0.15 to 0.3 inch long, usually black, tan, or black and tan

Distribution: Throughout North America

Comments: About a hundred species of the genus *Formica* have been described in North America. Most live in woodlands and fields, where they construct large mounds in relatively sunny places, as they require some sunlight for the health of the colonies. Most species of the genus *Formica* prey on other insects and are important in controlling some pests, especially tent caterpillars. Some species raid other colonies and bring workers back to their nest to act as slaves. Wood ants farm symbiotic aphids and treehoppers. The ants feed on the honeydew that the aphids or treehoppers produce and in return protect them from predators. When winter arrives, the ants take aphid and treehopper eggs to the safety of their nests and return the young aphid or treehopper nymphs to plants in the spring.

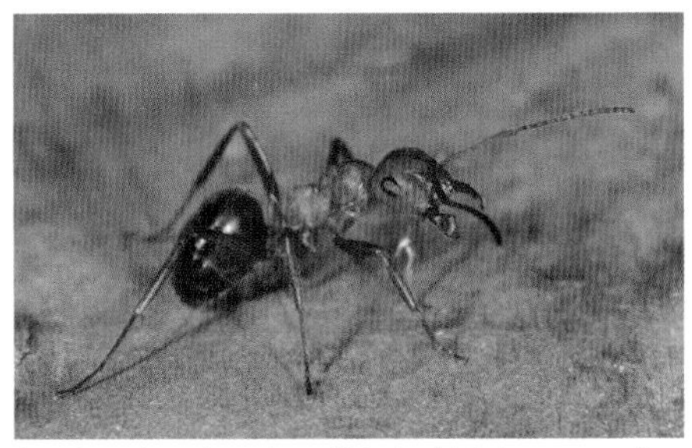

Ants have large jaws (mandibles), which they use to dig; carry earth, eggs, pupae, and food; and defend their nest. When they bite, some ants, such as fire ants, swing their abdomens around and squirt venomous compounds into the wound that they made with their mandibles.

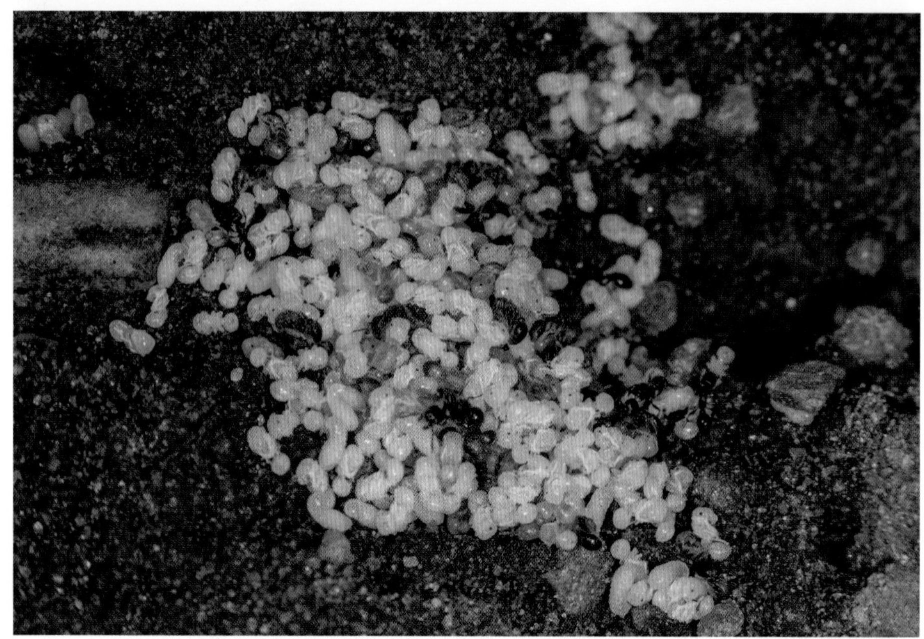

Ants are fond of heat. On sunny days they often move their eggs, larvae, or, in this case, pupae from their underground nest to under rocks that are warmed by the sun.

THIEF ANT, GREASE ANT

Latin name: *Solenopsis molesta*

Family: Formicidae

Identification: 0.03 to 0.1 inch long, reddish to brown, males black

Distribution: Throughout North America

Comments: Thief ants are so named because they often make their nest under rocks near the nests of other ant species so they can rob their food. Unlike most other ants, they build their nest just about anywhere. Because they are so small, they can live undetected between floorboards or cracks in homes and can even enter sealed packages of food. They are especially fond of grease and are often called grease ants. Their colonies usually range from a few hundred to a few thousand individuals and have several queens. *Solenopsis molesta* will eat just about anything but usually feed on fungal spores and dead animals. Like many other ants, when they find food, they leave a scent trail when they return to the colony for other ants to follow.

GLOSSARY

abdomen: third segment of the insect that contains the reproductive system and most of the digestive and respiratory systems.

aerenchyma: a soft plant tissue containing air spaces, found especially in many aquatic plants.

alates: winged reproductive ants and termites.

allomone: any chemical substance produced and released by an individual of one species that affects the behavior of a member of another species to the benefit of the originator but not the receiver.

antennae: primary sensory organs of insects.

cantharidin: a toxic, odorless blistering agent that is secreted by many species of blister beetles and related beetles.

carrion: the decaying flesh of dead animals.

cerci: sensory organ at the end of the abdomen of many insects.

chemoreceptors: a sensory cell or organ responsive to chemical stimuli.

chitins: polysaccharides (molecules composed of many sugars linked together) of the exoskeleton.

chrysalis: butterfly pupa.

circadian clock: the biological clock that regulates many physiological processes.

clypeus: a broad plate at the front of an insect's head.

colonial insects: insects that live in a highly organized social structure.

complete metamorphosis: where the life cycle includes four stages: egg, larva, pupa, and adult.

cornicles: backward-pointing tubes that project from the abdomen of aphids through which the aphid can secrete defensive molecules or pheromones.

cotyledon: an embryonic leaf in seed-bearing plants, one or more of which are the first leaves to appear from a germinating seed.

crepuscular: active at dawn and dusk.

daphnia: a small freshwater crustacean.

echolocation: detecting the location of objects by reflected sound; used by bats.

elytra: forewings of beetles.

exoskeleton: the rigid external support of the body of arthropods.

femur (plural: femora): the thigh and typically the longest segment of an insect's leg.

foregut: the portion of the digestive tract nearest the mouth.

forewings: the anterior wings of a four-winged insect.

fovea: a small area in the insect's eye where visual acuity is highest.

frass: the solid excrement of insects.

frons: the forehead or equivalent part of an animal, especially the middle part of an insect's face between the eyes.

derivative insects: more recently evolved insects.

desiccation: removal of water, to dry out.

diurnal: active during the day.

genal comb: a row of spines below the head of an insect.

head: first segment of the insect that contains the antennae, mouth, and eyes.

hemelytra: partially hardened forewing of certain insect orders including true bugs and grasshoppers.

hemolymph: insect blood.

hindgut: the portion of the digestive tract nearest the anus.

hindwings: the posterior wings of a four-winged insect.

holarctic: the name for the biogeographic realm that encompasses the majority of habitats found throughout the northern continents of the world.

holdfast: a stalked organ by which an alga or other simple aquatic plant or animal is attached to a substrate.

holometabolous: complete metamorphosis.

honeydew: a sugar-rich liquid secreted from the anus of hemipterans that feed on sap.

incomplete: simple morphogenesis where there is no pupal stage and the young look like the adults.

instar: a phase between two periods of molting in the development of an insect larva; e.g., the first install is the nymph or larva that emerges from the egg, and so on.

kleptoparasitism: a form of feeding in which one animal takes prey or other food from another animal that has caught, collected, or otherwise prepared the food, including stored food.

labium: a fused mouthpart that forms the lower lip of the mouth of an insect.

larva (plural: larvae): juvenile form of insects that undergoes complete metamorphosis.

Malpighian tubules: tubular excretory organs which open into the insect's midgut.

mandibles: pair of appendages near an insect's mouth used to grasp, crush, or cut food or in defense.

mesothorax: the middle of the three segments in the thorax.

metamorphosis: the process of transformation from an immature form to an adult.

metathorax: the last of the three segments in the thorax.

molting: the replacement of an old exoskeleton with a new, larger one.

naiads: the aquatic young of insects that undergo incomplete metamorphosis.

nymphs: the terrestrial young of insects that undergo incomplete metamorphosis.

ocellus (plural: ocelli): a simple eye of an insect.

omnivore: eats food of both plant and animal origin.
ommatidium (plural: ommatidia): each of the optical units that make up a compound eye.
osmeterium (plural: osmeteria): a protrusible glandular process of swallowtail larvae that emits a disagreeable odor for defensive purposes.
ovipositor: tubular organ through which female insects deposit their eggs.
ovoviviparous: producing live young from eggs that hatch within the body of the mother.
palpi: a pair of sensory appendages near the mouth of an insect.
parasitoid: an insect whose larvae live as parasites that eventually kill their hosts.
parthenogenetic: unfertilized ovum that develops into an individual that is identical to its mother.
pheromones: volatile hormones.
pollen basket: an area of hairs on the hind leg of bees used to carrying pollen.
prenatal comb: a row of spines behind the head.
proboscis: an elongated sucking mouthpart that is typically tubular and flexible in butterflies, moths, and certain flies or a sheath containing mouthparts in hemipterans.
prolegs: fleshy abdominal limbs of a caterpillar.
pronotum: sclerite that covers the insect's thorax.
prothorax: the first of the three segments in the thorax.
pterostigma: a dark pigmented spot on the leading (front) edge of the wings of dragonflies and some other insects that facilitates efficient gliding.
pubescent: hairy insect.
resilin: an elastic protein in insect joints.
rostrum: beak-like sheath containing stylets in hemiptera.
sclerites: individual pieces of the exoskeleton.
sclerotin: a structural protein that forms the cuticles of insects and is hardened and darkened by a natural tanning process in which protein chains are cross-linked by quinone group.
scopal: hairs for carrying pollen, on the underside of the abdomen of some species of bees and on the hind legs of other species.
setae: minute hairs connected to sensory nerves.
skeletonizer: a larva that feeds on the soft part of leaves while leaving the veins intact.
spiracles: valves at the opening of the trachea.
spermatophore: a protein capsule containing a mass of spermatozoa transferred during mating.
spermatozoa: male sex cell or gamete.
stridulate: to produce sound by rubbing the legs, wings, or other parts of the body together.
stylets: mouthparts of hemiptera.

subimago (subadult): winged stage in the development of mayflies before the adult stage.

thorax: middle segment of the insect that contains the wings and legs.

tibia: long, slender leg component below the femur.

trachea: breathing duct of insects, analogous to vertebrate lungs.

tracheoles: small, usually fluid-filled tubes at the terminus of the trachea.

tympanum: hearing organ in certain insects.

INDEX

A

Acanaloniidae, 74
Acrididae, 50–52
alderfles, 157
Alydidae, 92, 93
amberwing, eastern, 25
ambush bug, jagged, 83
Andrenidae, 282
angle-wing, lesser, 56
Anisopodidae, 222
ant lion, spotted, 154
Anthribidae, 109
ants, Dracula, 286
ants, field, 287
ants, grease, 288
ants, mound, 287
ants, thief, 288
ants, wood, 287
aphid, crapemyrtle, 66
aphid, goldenglow, 65
aphid, milkweed, 65
aphid, oleander, 65
Aphididae, 65–66
Aphrophoridae, 72
Apidae, 276–78
apollo, Rocky Mountain, 162
Aradidae, 98
Ascalaphidae, 158
Asilidae, 242–45
assassin bug, 79
Attelabidae, 107–8
Attevidae, 176

B

backswimmers, 85
Baetidae, 36
barklice, 16
bed bug, 97
bee hunters, 262
bee killers, 262
bees, bumble, 278
bees, cuckoo, 284
bees, digger, 282
bees, eastern carpenter, 279
bees, honey, 276–77
bees, leafcutter, 280
bees, mason, 280
bees, mining, 282
bees, resin, 281
bees, sweat, 283
bees, virescent green metallic halictid, 284
beetle, aquatic leaf, 119
beetle, American carrion, 143
beetle, American oil, 130
beetle, bark, 106
beetle, beach, 145
beetle, bostrichid, 131
beetle, case-bearing leaf, 113
beetles, checkered, 147
beetle, click, 148
beetle, disteniid longhorned, 124
beetle, dogbane, 113
beetles, dung, 140
beetle, false blister, 129
beetle, false potato 114
beetle, fire-colored, 133
beetle, flower longhorn, 125
beetle, fuller rose, 102
beetle, golden, 113
beetle, goldenrod soldier, 148
beetle, golden tortoise, 118
beetle, grapevine, 138
beetles, ground, 141
beetle, Hercules 139
beetle, imported long-horned, 104
beetle, jackknife, 148
beetle, Japanese, 136
beetles, jewel, 132
beetle, leaf-rolling, 108
beetle, margined carrion, 141
beetles, metallic wood-boring, 132
beetle, oriental, 137
beetles, pintail, 144
beetle, powder-post, 131
beetles, predaceous diving, 149
beetles, predaceous water, 149
beetle, red milkweed, 127
beetle, rhinoceros, 139
beetles, ripiphorid, 145
beetles, rove, 150
beetle, scooped scarab, 140
beetle, spotted cucumber, 116
beetle, skeletonizing leaf, 115
beetle, striped cucumber, 122
beetle, sumac flea, 117
beetle, thistle tortoise, 118
beetles, tiger, 123

beetle, tomentose
burying, 142
beetle, trogossitid, 144
beetles, tumbling
flower, 144
beetle, viburnum leaf, 120
beetle, warty leaf, 121
beetles, wedge-shaped, 145
beetles, whirligig, 146
beetle, zebra longhorn, 126
Belostomatidae, 89
Berothidae, 153
Berytidae, 90
Bibionidae, 224
Blaberidae, 43
black damsel bugs, 91
black jacket, 256
black quill, 35
blue corporal, 26
Bombyliidae, 230–31
booklice, 16
borer, broad-necked
root, 124
borer, emerald ash, 132
borer, locust, 129
borer, red oak, 128
borer, red-necked cane, 133
Bostrichidae, 131
boxelder bug, 78
Braconidae, 274
branch pruner, 126
broad-headed bugs, 92
Buprestidae, 132–33
buckeye, 171
bumblebees, 278
butterfly, cabbage
white, 163
butterfly, monarch, 170
butterfly, zebra, 174

C

cabbage white, small, 163
caddisflies, 39
Calliphoridae, 225
Calopterygidae, 30
Cantharidae, 148
capsid bugs, 76
Carabidae, 123, 141
caterpillar, American
lady, 173
caterpillar, eastern
tent, 196
caterpillar, masked
birch, 212
caterpillar, red-fringed
emerald, 193
Crabronidae, 261–62
Cerambycidae, 124–29
Ceratopogonidae, 221
Cercopidae, 70–71
checkerspot, Baltimore, 168
chinch bug, saltmarsh, 77
Chrysididae, 268
Chrysomelidae, 113–122
Chrysopidae, 152
Chironomidae, 220
cicada, dog day, 67
cicada, water 86
Cicadellidae, 70–71
Cicadidae, 67
Cimicidae, 97
clearwing, snowberry, 181
Cleridae, 147
Coccinellidae, 110–12
cockroach, spotted
Mediterranean, 42
cockroach, dusty, 41
cockroach, Madagascar
hissing, 43
Coenagrionidae, 31–32
commodore, 171
Conopidae, 249–50
Coreidae, 94–96
Corixidae, 86
Corydalidae, 155–56
Crambidae, 197–98
cricket, cave, 59
cricket, fall field, 60
cricket, field, 60
cricket, house, 60
cricket, marsh ground, 61
cricket, snowy tree, 62
cricket, sphagnum, 61
cricket, sphagnum
ground, 61
cricket, spotted camel, 59
cricket, thermometer, 62
cricket, two-spotted tree, 62
Culicidae, 214–16
curculio, apple 103
curculio, rose, 107
Curculionidae, 101–6
Cynipidae, 270

D

damselflies, spreadwing, 30
dobsonfly, 155
Dolichopodidae, 237
doodlebug, 154
Drepanidae, 212
dun, chocolate, 35
dun, speckled, 36
Dytiscidae, 149

E

earwig, European, 6
Ectobiidae, 41–42
Elateridae, 148

Empididae, 238
Entomobryidae, 1, 2
Ephemerellidae, 35
Erebidae, 182–87
Eupelmidae, 263

F

firefly, 134
firefly, winter, 135
fishflies, 156
fishflies, spring, 156
fishmoth, 4
Flatidae, 73
flat bugs, 98
flat bark bugs, 98
flea, cat, 14
flea, oriental rat, 14
flea, rat, 14
flea, tropical rat 14
flies, assassin, 242
flies, bee, 230
flies, bee-like robber, 243
flies, black-tailed bee, 231
flies, crane, 218
flies, drain, 223
flies, drone, 235
flies, gnat-ogres, 244
flies, greenhead, 228
flies, green bottle, 225
flies, goldenrod gall, 247
flies, flesh, 250
flies, flower, 235
flies, four-barred knapweed gall, 229
flies, fruit, 230
flies, hover, 235
flies, long-legged, 237
flies, long-tailed dance, 238
flies, march, 224
flies, marsh, 229
flies, moth, 223
flies, narcissus bulb, 237
flies, phantom crane, 219
flies, picture-winged, 239
flies, robber, 242
flies, seedhead, 229
flies, snail-killing, 229
flies, stilt-legged, 234
flies, soldier, 241
flies, stiletto, 246
flies, syrphid, 235
flies, tachinid, 232–33
flies, thick-headed, 249
flies, thread-waisted robber, 245
flies, tiger bee, 231
flies, transverse flower, 234
flies, winter crane, 219
Forficulidae, 5, 6
forktail, citrine, 32
Formicidae, 286–88
froghopper, meadow, 72
froghopper, red-legged, 71
froghopper, two-lined, 70
Fulgoridae, 99

G

gadflies, 227
gnat, dark-winged fungus, 217
gnat, fungus, 217
gnat, window, 222
gnat, wood, 222
gnats, 220
Gerridae, 88, 89
Geometridae, 190–95
grasshopper, grizzly spur-throated, 51
grasshoppers, grouse, 53
grasshopper, long-horned, 57
grasshopper, northern green-striped, 52
grasshopper, pine tree spur-throated, 51
grasshoppers, pigmy, 53
grasshopper, two-striped, 50
grass bugs, 76
Gryllidae, 60–62
Gyrinidae, 146

H

hairstreak, eastern tailed-blue, 165
Halictidae, 283–84
heliconian, zebra, 174
Hesperiidae, 161
hornet, bald-faced, 256
hornet, white-tailed, 256
horseflies, 227

I

Ichneumonidae, 271–73

J

Jesus bugs, 87
jewelwing, ebony, 30

K

katydid, broad-winged, 56
katydid, forked-tailed, 56
katydid, forked-tailed bush, 56
katydid, meadow, 57
katydid, northern, 55
katydid, northern true, 55

katydid, true, 55

L

lacewing, antlion, 154
lacewing, beaded, 153
lacewing, brown, 152
lacewing, green, 153
ladybird, cream-spot, 111
ladybird, polkadot, 111
lady beetle, Asian 110
lady beetle, Asian multicolored, 110
lady beetle, fourteen-spotted, 111
lady beetle, harlequin, 110
lady beetle, hyperaspis, 112
Lampyridae, 134–35
lanternfly, spotted, 99
Lasiocampidae, 196
leaf bugs, 76
leaf-footed bugs, 96
leaf-footed bugs, distinct, 95
leafhopper, candy-striped, 70
leatherwing, Pennsylvania, 148
Lepismatidae, 4
Leptophlebiidae, 35
Lestidae, 30
Libellulidae, 24–27
lice, canine chewing, 12
lice, crab 11
lice, dog, 12
lice, jumping plant, 18
lice, human, 11
lice, pubic, 11
lightning bug, 134
Limacodidae, 209
Limnephilidae, 38
long-wing, zebra, 174
looper, lesser grapevine, 190
lupine bug, 93
Lycaenidae, 165
Lygaeidae, 77

M

Mantidae, 45–46
mantis, Chinese, 45
mantis, European, 46
mayfly, prong-gilled, 35
Megachilidae, 280–81
Meloidae, 130
Membracidae, 68–69
Micropezidae, 234
midges, 220
midges, biting, 221
midges, green, 220
miner, goldenrod leaf, 119
Miridae, 75–76
Mordellidae, 144
mosquito, 215
mosquito, Asian tiger, 216
mosquito, forest, 216
mosquito, tiger, 216
mosquito larvae and pupae, 214
moth, arched hooktip, 212
moth, azalea hawk, 177
moth, azalea sphinx, 177
moth, banded tussock, 182
moth, cecropia, 188
moth, crocus geometer, 191
moth, false crocus, 191
moth, grapeleaf skeletonizer, 200
moth, hawk, 179
moth, Hebrew, 206
moth, hummingbird, 180
moth, io, 189
moth, Isabella tiger, 184
moth, lapped, 196
moth, grape leaf folder, 197
moth, gypsy, 183
moth, joker, 205
moth, jocose sallow. 205
moth, orange-patched smoky, 199
moth, pale tiger, 182
moths, plume, 211
moth, primrose, 208
moth, roadside sallow, 207
moth, silk, 204
moth, small-eyed sphinx, 179
moth, snowy urola, 198
moth, sphinx, 179
moth, spiny oak slug, 209
moth, suzuk's promalactis, 210
moth, turbulent phosphila, 202
moths, underwing, 185
moth, virgin tiger, 186
moth, white-dotted prominent, 201
moth, winter, 195
moth, woolly bear, 184
moth, yellow-collared scape, 187
Mutillidae, 269
Myrmeleontidae, 154

N

Nabidae, 91
Naucoridae, 84

naiads, damselfly, 29
naiads, dragonfly, 24
naiads, mayfly, 34
naiads, tube-making caddisfly, 38
needle bugs, 84
Nepidae, 84
Noctuidae, 205–8
no-see-ums, 221
Notonectidae, 85
Nymphalidae, 168–74

O

Oecophoridae, 210
Oedemeridae, 129
orange bluet, 31
owlflies, 158

P

pale beauty, 194
Panorpidae, 22
pansy, 167
Papilionidae, 166–67
parnassian, Rocky Mountain, 162
partridge bugs, 75
partridge scolops, 75
pearl crescent, 169
Pediculidae, 11
Pentatomidae, 80–82
Phasmidae, 48
Pieridae, 163
planthopper, citrus flatid, 73
planthopper, two-striped 74
plant bugs, 76
plant bug, meadow, 75
praying mantis, 46
praying mantis, Chinese, 45
psocids, 18
Psocoptera, 16
Psyllidae, 18
Psychodidae, 223
Pterophoridae, 211
Ptychopteridae, 219
psyllids, 18
Pulicidae, 14
punkies, 221
Pyrochroidae, 133

R

Raphidiidae, 159
Reduviidae, 79, 83
red admiral, 172
Rhaphidophoridae, 59
Rhinotermitidae, 9
Rhopalidae, 78
Rhyparochromidae , 80
Ripiphoridae, 145
roach, wood, 42

S

saddlebags, 27
saddlebag gliders, 27
Saldidae, 76
Saturniidae, 188–89
saucer bugs, 84
sawflies, 254
sawfly larvae, common, 254
sawfly, nothern, 254
Sciaridae, 217
Sciomyzidae, 229
Scarabaeidae, 136–40
seed bug, long-necked, 80
seed bug, western conifer, 94
scorpionflies, 22
sharpshooter, broad-headed, 72
shieldbacks, eastern, 54
shield bugs, 82
shore bugs, 76
Sialidae, 157
silkworm, 204
Silphidae, 141–43
silverfish, 4
skipjack, 148
skippers, 161
small milkweed bug, 77
snakeflies, 159
snout-moth, 198
spanworm, currant, 192
Sphecidae, 265–67
Sphingidae, 177–81
sphinx, grapevine, 178
sphinx, hog, 178
sphinx, Virginia creeper, 178
spider cricket, 59
spinner, early brown, 35
spittlebug, meadow, 72
spittlebug, red-legged, 71
spittlebug, two-lined, 70
spreadings, 30
springtails, slender 2
stick, northern walking, 48
stick insect, 48
stick insects, water, 84
stilt bug, spined, 90
stink bug, brown, 81
stink bug, brown marmorated, 98
stink bug, green, 80
stink bug, predatory, 81
stink bug, rough, 82
stink bug, two-spotted, 81
sulphur, clouded, 164
sulphur, common, 164
swallowtail, American, 166

swallowtail, black, 166
swallowtail, spicebush, 167
Syrphidae, 234–37, 248

T

Tabanidae, 226–28
Tachinidae, 232–33
termite, eastern
subterranean, 8
Tenebrionidae, 145
Tenthredinidae, 253–54
Tephritidae, 229–30, 247
Tetrigidae, 53
Tettigoniidae, 54–57
Therevidae, 246
Thripidae, 20
thrips, 20
Tipulidae, 218
toe-biters, 89
treehoppers, buffalo, 69
treehopper, thorn-mimic, 68
Trogossitidae, 144
Trichoceridae, 219
Trichodectidae, 12
true water bugs, 149

U

Ulidiidae, 239–40

V

Vespidae, 256–60

W

wasp, bee hunter, 262
wasp, braconid, 274
wasp, chalcid, 264
wasp, cuckoo, 268
wasp, ectemnius, 267
wasp, ensign, 267
wasp, European paper, 258
wasp, gold, 268
wasp, grass-carrying, 266
wasp, hatchet, 267
wasp, horntail, 255
wasp, hunting, 265
wasp, ichneumon, 271
wasp, mason, 259
wasp, nightshade, 267
wasp, oak rough bullet
gall, 270
wasp, pigeon horntail, 255
wasp, pigeon tremex, 255
wasp, potter, 259
wasp, sand, 261
wasp, short-tailed
ichneumon, 273
wasp, thread-waisted, 265
wasp, tiphiid, 264
wasp, weevil-hunting, 262
wasp, velvet ant, 269
waterbees, 85
water boatmen, 86
water bugs, creeping, 84
water bugs, giant, 89
water scorpions, 84
water striders, 88
webworm, ailanthus, 176
weevils, acorn 101
weevil, broad-nosed, 105
weevil, fuller rose, 102
weevil, fungus, 109
whitetail, common, 25
white wax ladies, 112

X

Xylocopinae, 279

Y

yellowjacket, eastern, 257

Z

Zygaenidae, 199–200

ABOUT THE AUTHOR

David M. Phillips has had an interest in insects and photography since he was a teenager. His doctoral thesis at the University of Chicago involved electron microscopy of the reproduction of a fungus gnat. As a postdoctoral fellow at Harvard Medical School, he pursued the study of sperm function in insects. This involved collecting and identifying hundreds of insect species. Although he continued to work on insects for a number of years, the majority of his research involved mammalian reproduction. In the mid-1980s, with the emergence of HIV and AIDS, Phillips set aside his research on the reproduction of insects and mammals to use his skills as a biologist to study HIV and AIDS transmission and strategies for prevention. This effort continued until his retirement in 2008.

Phillips has published more than 200 scientific papers and hundreds of his photographs and electron micrographs appear in textbooks and other scientific and educational materials. He has always loved to teach young people and has taught at Harvard Medical School, Washington University, New York Medical College, SUNY Downstate College of Medicine, and Weill Cornell Medical School.

In the 10 years since retirement, Phillips has returned to the study of insects, first writing *Art and Architecture of Insects* (University Press of New England), which is illustrated with scanning electron micrographs that he had taken over the years, and more recently concentrating on macrophotography of insects. His activities in three photography clubs on Cape Cod has helped him keep up with the latest advances in digital photography. He lives in Yarmouth Port, Massachusetts.